MORALITY —
NATURE'S CROWNING ACHIEVEMENT
The Making of our Moral Compass

BILL WILSON

authorHOUSE®

AuthorHouse™ UK
1663 Liberty Drive
Bloomington, IN 47403 USA
www.authorhouse.co.uk
Phone: UK TFN: 0800 0148641 (Toll Free inside the UK)
*   UK Local: (02) 0369 56322 (+44 20 3695 6322 from outside the UK)*

Published by AuthorHouse 12/06/2022

ISBN: 978-1-7283-7660-8 (sc)
ISBN: 978-1-7283-7661-5 (e)

Library of Congress Control Number: 2022921181

Print information available on the last page.

This book is printed on acid-free paper.

Scripture quotations are taken from The Message. Copyright 1993, 1994, 1995, 1996, 2000, 2001, 2002. Used by permission of NavPress Publishing Group.

Without civic morality communities perish, without personal morality their survival has no value.
Bertrand Russell, *Authority and the Individual*

DEDICATION

To my wife, Rita, who has been a constant source of encouragement and a perceptive contributor to my reflections and writings.

ACKNOWLEDGEMENTS

To those many people whom I have met along the way
who have caused me to reflect and adjust.
To Kieran, for his guiding hand in all matters digital.
To Rita, my wife, for thoughtfully and
painstakingly reviewing every draft.
To my good friend Henry, for his insightful comments.
To all the team at AuthorHouse, especially Homer for his
encouragement and guidance through the publishing process.
To you all, my grateful thanks.

CONTENTS

PREFACE

This book is not an academic treatise but simply the product of sixty years of sailing the sea of life, reading, thinking, conversing, and engaging with family, friends, and colleagues. I am not a scholar, nor an expert in neuroscience, or psychology, or philosophy, or theology, but simply an ordinary person who has studied and practiced professionally in a number of fields, and who has felt a constant compulsion to seriously reflect upon significance and meaning as events have impacted upon me day after day. I find myself constantly driven by the words of Socrates: 'An unexamined life is not worth living.'

I long ago became fascinated by the question 'Who am I?'

For many years, in trying to find the answer to this question, I simply followed the model of thinking of nonconformist Christianity bequeathed to me by my illustrious forefathers (many of them founders and leaders of the Primitive Methodist Church in the north of England) Gradually, however, through following the path of reason into the world of science via the world of engineering, I found that a 'bottom-up' approach to understanding the meaning of life was far more satisfying to my inner being than the 'top-down' revelation model which cocooned my forefathers. This change was symbolized for me by the increasing use in ordinary society of the word *psychological* and the decreasing use of the word *spiritual* in everyday conversations. I interpreted this, not as a denigration of the non-physiological dimension of human life, but rather to signify a more meaningful, rational understanding of it. And with such an understanding, the realization that the responsibility for achieving fulfilment and enrichment of our lives lay within our own hands as developing human beings and human societies, and not in obedience to the dictates of a supernatural power.

This book is about my personal journey through these complex issues.

In every field, no matter the scholarly depth of enquiry, or the razor-sharp focus on minutiae, ultimately the participant must find some way to live out authentically the ordinary routines of life. In arriving at this necessary accommodation of everyday life, some purity of understanding – from the expert's point of view – may be lost. So be it. Nevertheless, despite the absence of purity or perfection, the living practice of life will have been enhanced. This is the wellspring of this book.

From such an encounter with life, over many years and embracing a number of fields, I have come to an understanding of human development as an organic process, part physiological, part psychological (neither of which is exclusive of the other): from the moment of conception cells begin to divide and interact and build up electrochemical reactions which culminate, usually in sequence, in the emergence of a self- regulating, self-aware bundle of energy; this bundle of electrochemical energy then, amazingly, begins to probe for an understanding of itself and the impact upon it of the world around it. There are two fundamental characteristics of this new 'being' – a determination to survive (normally expressed as the 'will to live'), and an insatiable curiosity (at first expressed through engaging the five developing senses, and later through language, asking 'why?' about everything encountered in the world around it). This extraordinary 'being' presently stands at the apex of the whole evolutionary structure of nature, and I cannot but stand in awe of it!

The most critical term in the above paragraph is *self-aware*, for it expresses a state of being which as yet defies explanation. As Sam Harris says in his inspiring book *Waking Up*:

> Whatever the ultimate relationship between consciousness and matter, almost everyone will agree that at some point in the development of complex organisms like ourselves, consciousness *seems* to emerge. This emergence does not depend on a change of materials, for you and I are built of the same atoms as a fern or a ham sandwich. Instead, the birth of consciousness must be the result of organization: Arranging atoms in certain ways appears to bring about an experience of *being* that very collection of atoms. This

is undoubtedly one of the deepest mysteries given to us to contemplate.

The purpose of this book is the exploration not of this mystery but rather the emergence of morality within consciousness once in being. It therefore explores the interplay between the two characteristics of 'being' mentioned above, the physiological and psychological, and the resultant recognition that cooperation with the 'other', that is, everything outside the 'self', is the best way not only to secure basic survival but also to satisfy the higher desires now released. It is the satisfaction of this non-physiological (one might say spiritual) characteristic, which I believe to be of fundamental importance to a deep sense of fulfilment in the inner being.

At our birth then, we are primarily a bundle of electrochemical energy but with these two extraordinary defining attributes: a mechanism which enables us to evaluate how best to preserve that energy, and an enquiring mind intent on exploring every aspect of 'being', both within and without. Initially, along with the rest of the animal kingdom, preserving that energy simply means finding sustenance for the body. Gradually, however, as that existential need is satisfied, our enquiring mind asks more and more searching questions and begins to evaluate not only the best way to secure its physical survival, but also how it can best satisfy developing inner longings for a sense of fulfilment, contentment, and peace. It is as if we need to find vehicles which will utilize the electrochemical energy of our being so perfectly that we become truly harmonized beings: we need to be plugged in, as it were, to the elemental essence of all life, for our experience of being to become truly authentic. In this process our relationship to the 'other' is of utmost importance.

Our insatiable curiosity, our compulsion to reason, to ask the question 'why?' which arises within us almost from the day we are conceived – certainly from the dawn of our consciousness of self – has brought us huge benefits. It has led to a continuous enrichment of our experience of life through a fuller understanding of ourselves and the world around us. It has conquered many of our diseases. It has overcome many grievous obstacles. It has greatly expanded our horizons. It has brought us, and continues to bring us, insight into the very intricate web of life itself. With such insight the humdrum, the mundane, the animalistic, the ordinary, is transformed into the extraordinary, the amazing, the profound. Life becomes 'holy' only in its totality, and we cannot but stand in awe of it. From time to time our souls are uplifted and enhanced by our glimpse of the unseen agent responsible for all these transformations – love. We become aware that, for the human species, it is *love* which holds everything together – not hate, not envy, not pride, not selfishness, not power, not status, not possessions, only love; that our own sense of fulfilment and happiness is inextricably bound up in the happiness of the 'other'.

Our curiosity then, gives rise to an evaluation of that which is 'good' for us and that which is 'bad'. And gradually we come to recognize that having a concern for the 'other' outside ourselves is what ultimately brings us total contentment within our inner being. And the 'other' is everything that exists outside our own consciousness of self – other human beings, the natural world, literature, art, music. There is a unity, then, in our experience of the whole world of existence. No separations, no divisions, only unity.

Consciousness, then, expressed in the life we live every day, is an all-pervading experience of complexity, and only an agreed code of cooperation with the 'other' – that is, that which is other than the self – will help us navigate our way through and enjoy some measure of human flourishing.

There are many perilous rocks and hidden sandbanks in the sea of life, which threaten our happiness, our stability, and our survival. Therefore, we need a robust moral sensibility to guide us. The domain of morality is integral to our nature, but its development is something we must nurture. Morality is at the apex of human development, which is why Bertrand Russel said that without it we perish.

This little book traces the fruits for humanity of following the path of reason, explores the making of a strong moral identity, and tentatively presents, in diagrammatical form, the constituents of a moral compass. Finally, it suggests practical ways in which our moral sensibility can be developed and nourished.

INTRODUCTION

The Pace of Change

As the amazing products of the eighteenth-century Enlightenment have now trickled down to us ordinary citizens – in the fridges and freezers which help us preserve our food; the new hips, knees and hearts which keep us upright and active for many more years; the mobile phones and i-devices which keep us in instantaneous touch with anyone anywhere in the world and which put at our fingertips unlimited information – we have a level of understanding of the world that even our grandparents could never have imagined, let alone our earlier ancestors!

We move from place to place independently, in our own private motor cars; we fly at the drop of a hat, to any part of the world in a matter of hours; and we are planning at this very moment to colonize another planet. Wow! Whatever will be the kind of world our children inhabit?

Following the Path of Reason

All these changes in our world have come about by following the path of the Enlightenment, the path of reason: this path demands reflecting deeply upon the evidences surrounding us; asking the question 'why?' at every turn; and being constantly open to change in the light of new evidence, no matter what the cost to our existing understandings and resultant ways of life.

Rejecting 'God', but Holding On to Morality

Following the path of reason, it is not surprising that our understandings of the world today are markedly different from those of our forefathers – in all matters except, it seems, religion. The official teachings of the Christian Church have altered little in the past two thousand years. Not surprisingly, therefore, belief in the God it portrays has declined significantly. A YouGov survey in the summer of 2021 revealed that over 50 per cent of us believed in aliens, while a similar survey in 2020 revealed that only 43 per cent of us believed in a god or higher spiritual power. Further, according to the British Social Attitudes survey for 2018, only 1 per cent of people aged 18–24 identify as Church of England, and in all groups, people identifying as Christian has fallen to 38 per cent from 66 per cent in the same survey in 1983.

Yet morality, which traditionally was believed to be God's rules for human living, both for the individual and society, is still universally acknowledged to be essential to human flourishing and indeed to human survival.

How have we arrived at this belief in morality without God?

The Unstoppable Question

We human beings seem unable to stop asking the question 'why?' Almost from the day we are born, we question things; it seems wired into our brains that in every moment of every day we ask the question 'why?' It might be about plants or insects, stars or planets, activities or desires, illness or tiredness, ingrowing toenails, or hair loss! Every aspect of our lives is probed by the unstoppable question 'why?' It is obviously a question built into our very state of being. It is as essential to us as breathing. It is the product of an *enquiring mind*.

And it seems that the human species has a superabundance of 'enquiring mind' over and above the rest of the animal kingdom. It is this enquiring mind which has propelled the human species into worlds unknown to the rest of the natural world. It is therefore an attribute to be greatly prized.

The most fundamental 'whys', the most important 'whys' promoted by this enquiring mind, are the 'whys' of 'Why are we here?' 'Why do we have our being?' 'What are we supposed to get out of life?' 'Is there a purpose to life?'

The Most Important Question

The answers to these profound, existential questions, are dependent on the answer to an even more primal question: 'Where did we come from?' 'How did we get here?' 'How did we come to be?'

– The historical answer to this question has been, for the human species across the world, that a Higher Power created us. Therefore, as we gradually become aware of each dimension of our humanity – life, death, love, hate, happiness, sadness, elation, contentment, meaning, purpose – we must ask, of every experience: 'What does our Creator expect of us?' In other words, we strive to find and follow our Creator's instructions on everything. Any moral code which helps to keep us on the 'straight and narrow' is, therefore, the moral code which our Creator has provided.

– The contemporary answer to the question 'How did we come to be?' is, however, somewhat different. Following the path of the constantly enquiring mind has led us to understanding the world as a process of evolution. And the sustaining energy of this process is the inner 'will to live'. We have come to understand that everything survives because each species has found an environment conducive to its own unique 'will to live'. For the animal kingdom, this has involved, in response to the individual existential requirement to constantly replenish energy, practicing the principle of the survival of the fittest. This is how, in evolutionary understanding, all nature sustains itself.

The human species has evolved within this system, but through a highly developed self-consciousness, incorporating an enquiring mind, the human species is now able to satisfy the raw will to live with abundance and with a minimum of violence, and now seeks non-material (spiritual) satisfactions.

The Necessity for Morality to Human Flourishing

Within this understanding of life, a moral code serves to enhance human flourishing in every dimension. It promotes the moderation, if not elimination, of the concept of the survival of the fittest, and it helps bring about deep, inner, spiritual satisfaction. Of course, this still takes place within the fundamental requirement for self-survival; the will to live is still paramount and must be fulfilled. But alongside this, other less physiological more psychological (spiritual) needs claim our attention.

The interplay between these two drives, as they are sometimes called, has led many scholars, including John Stuart Mill, John Locke, Adam Smith, and Friedrich von Hayek, to accept the concept, first proposed by Alexis de Tocqueville in his work *Democracy in America* (1835), that all human actions, moral actions included, are motivated, at best, by *enlightened self-interest*, that is, that by serving the interests of others, individuals are, in many instances, finding the best way of serving themselves.

It certainly would appear, from the history of humankind, that only a very few exceptional human beings, and in very particular circumstances, have overcome the raw power of the will to live and the subtler drive of enlightened self-interest, and offered their very lives in the cause of a greater (spiritual) good – as in the case of a hunger striker, or a mother who gives her life to save the life of her child.

Most of us, however, having evolved to an awareness of a moral dimension to life, are still, even at the highest level, not able to completely free ourselves from a degree of enlightened self-interest, even though we perceive, if only dimly on the horizon, that our greatest satisfaction lies in contentment of the spirit, which ultimately may demand the sacrifice of the self.

The Olympian – Motivated by 'Spiritual' Satisfaction

A good example of this progress in the recognition of the role of the inner spirit in human motivations is the modern-day Olympian. The original motivation for developing athletic prowess has its roots in basic survival – the physical agility and prowess was a reaction to a perceived threat motivated by the existential will to live. However, alongside this it was soon recognized that the exercise of such prowess – of being able to run faster, throw farther, jump higher, lift heavier – gave rise, upon survival, to very satisfying feelings of achievement. Eventually, these feelings came to be sought after purely for themselves, outside the need to survive.

This sense of achievement is a feeling, a sensation, valued purely for itself, and it contributes greatly to human flourishing, to the sense of a achieve their emancipation. They all required the heroic, sustained intervention of *individuals* with broad understandings and deep empathy – in other words, individuals with a highly developed moral compass.

If we have not been created by a Supreme Being, who presumably would have created us for their own reason or purpose, then any reason or purpose we do discern must be of our own making, from our own assessment of the values we have placed on our experiences of the various and complex aspects of human existence. And, as we each individually evolve in our own historical, material world, we will each determine our own purpose. And this purpose will compete for our attention alongside our innate will to live. For particular individuals this discernment can be so strong that it overcomes even the almost invincible will to live. Such individuals bring about great changes in society – as in the examples above of the freeing of the slaves, the emancipation of women, the ending of racial and religious discrimination, and the acceptance of all forms of sexual differences. To such individuals all humanity owes a great debt.

Summary: Enormous Material and Spiritual Benefits Flowing from Our Embracing of Evolution

It is indisputable that following the path of science has brought huge material benefits to the human species: from the first accidental discovery that friction caused heat and fire and thus enabled food to be cooked and dwellings to be warmed; to the discovery of the lever and the wheel, enabling previously impossible loads to be moved and transported; to the discovery of chemical processes which alleviated pain and cured diseases. These and countless other discoveries are testimony to the enormous benefits the human species has enjoyed by following its enquiring mind.

It is also indisputable that it is not only in the outer, material, physical world that following the enquiring mind has so enriched human life, but also in the inner, spiritual world of the amazing experience of consciousness; in the search of the self for meaning and enrichment. Almost all of humankind now recognizes the role played by evolution in explaining this spiritual aspect of human development in all its depth and complexity. The universal need for companionship, for self-esteem, for self-fulfilment, motivates the greater part of our living experience today.

How Does This Work in Practice?

One of the finest researchers in this field was the American psychologist, A. H. Maslow. In 1943 he proposed a theory for understanding the psychological motivations behind a supreme feeling of satisfaction within the human soul, which all human beings feel from time to time and to one degree or another. I use the word 'soul' to refer to that deep inner centre of human awareness; the home, as it were, of my true self, of who *I* really *am*; the amazing sensation of the spiritual, emotional *me*, held within my material frame. It is the living practice of that me, my soul, in the world, which reveals my motivations, my morality, my basic evaluations of everything that impinges upon my existence. Day after day, hour after hour, countless moments of experience require my instant evaluation, before motivating my response. Some experiences are very basic – for example, the experience of hunger, which motivates my straightforward response of the search for food. But other experiences are much more subtle and intangible, such as my need for companionship and friendship, and the projection of imagination.

Maslow proposed that total human well-being depended on the satisfaction, in priority order, of *five innate human needs*. At the base of his hierarchy lay the physiological needs (for food, shelter, clothing); next came the safety needs (for surplus and security). These two sets of needs together formed the Basic Needs necessary for survival. Then came the Higher Needs, a group of three psychological needs: social needs (for company, friendship, love); ego needs (for status and self-esteem); and finally, self-actualization needs (for self-fulfilment and personal, spiritual, growth). The motivations enlivened to satisfy the first two levels of need on Maslow's Hierarchy of Needs – the Basic Needs necessary for survival – are very straightforward; the complexities arise, in ascending order, with the Higher Needs of social, ego, and self-actualization.

I am personally still of the view that Maslow's Theory of Motivation offers convincing explanations for many aspects of human behaviour. For example, it seems to me to offer a plausible explanation, particularly as modified by his later work, on the recent rise in popularism in the USA and Europe.

Due to the entry, in the latter half of the twentieth century, of Eastern and Far-Eastern low-cost manufacturing countries into the world economy, coupled at the same time with huge advances in sea transportation enabling large quantities of goods to be shipped cheaply anywhere in the world, the low-skilled Western industrial workers have either lost their jobs or had to take significant cuts in their wages and hence their standard of living. Following Maslow's theory, they have returned to the first and second levels on the hierarchy; that is to say, their prime motivation now is simply to survive – to put enough food on the table and a roof over their heads, and, if possible, to save a little extra for a treat now and again.

Furthermore, the original industrial revolution not only provided a much higher material standard of living, but also promoted strong social cohesion among those working in mass-labour industries: their native individuality was now complemented by a strong sense of togetherness – they worked together, lived together, holidayed together, socialized together. This 'combining' brought prosperity to their industry, better working conditions and practices for all, and a strong feeling of comradeship and well-being.

Now, following the global changes outlined above, they are once again alone: individual entities with little corporate identity; their communities fragmented; their industries – textiles, shipbuilding, steelmaking, coal mining – either long gone or reduced to a shadow of their former size, or, at best, replaced by much smaller, less labour-intensive industries, generally paying much lower wages. Individualism, rather than collectivism, is the current mood music. In this environment, for this section of society, it is not surprising that the higher needs of Maslow's hierarchy have little pull: the all-consuming motivation is simply to survive, and perhaps, with a little bit of luck, to achieve a small surplus.

In such circumstances, when people are only just managing to keep the boat afloat, they must give a wide berth to anything that might possibly rock the boat. Anything that will threaten the very tenuous hold on survival experienced by millions of people throughout the world will be quickly dismissed, and any radical change will be greatly feared. Ironically, this is also true for those at the opposite end of the scale of personal wealth: those who are living fulfilling lives at the top of Maslow's hierarchy, also fear change. The current arrangements are serving them well. Indeed, so long as nothing changes, they will do even better. Consequently, the gap between the rich and the poor, the 'haves' and the 'have nots', the very rich and the 'just about managing', grows wider in societies across the world.

It must be of concern to all of us that large swathes of the world's population have little opportunity to flourish as human beings because so much of their energy is expended on the immediate matter of survival. Of course, in evolutionary terms, the preservation of the species through continuously satisfying the 'will to live' is of paramount importance. But we, as a species, have also come to recognize that mere survival is not enough; we wish to fulfil *all* our potential, to enjoy the full flourishing of all the possibilities of our humanity.

The Need for Cooperation with Others and with the Environment

Studies in human anthropology now indicate that achieving our full potential is best achieved through cooperation rather than the relentless pursuit of naked, crude self-interest. Dr. Oliver Scott Curry said in the Humanists UK Darwin Day Lecture 2021, 'We, as the human species, have evolved to develop a collection of cooperative rules which helps us get along together, and promotes the common good.'

For the human species to do well and to flourish, then, we need to cooperate with others and with our environment, to 'love our neighbours as ourselves', as Jesus of Nazareth put it. Our flourishing will not be achieved through the pursuit of brutal self-interest, but – at least for most of us – through enlightened self-interest, embracing not only the first two needs of Maslow's hierarchy but also the higher three. The human species flourishes and advances only when every member works together. This cooperative environment promotes human contentment of the highest order.

The Urgent Need for a Moral Compass

To work in concert with others day by day, we need to have something 'second nature' within us that will help us make good instantaneous decisions in very complex situations: a naturally functioning moral compass. This book sets out to show that morality is indeed a natural feature of humanity: that it arises within us with the dawn of our consciousness, but, as with all other aspects of our beings, it needs to be constantly nurtured if we are to achieve our full human potential. In short, we need to 'grow' a moral compass within our souls, at the very heart of who we sense ourselves to be.

And the need is urgent.

The world is standing on the brink of an unparalleled new age; an age in which unlimited intelligence will be unleashed to act on all human affairs; where digital technology will infiltrate every aspect of our existence, reaching into the very core of our humanity; where our world's resources are being depleted at an alarming rate, where our planet is suffocating from the by-products of our activities.

In such a world, our moral awareness must be comprehensively alert and finely tuned. Without it, the world will descend into chaos and darkness.

PART ONE

CHANGING TIMES

Today we are living healthier, wealthier lives – and it's
thanks to the values of the Enlightenment ....
David Aaronovitch, *The Times*

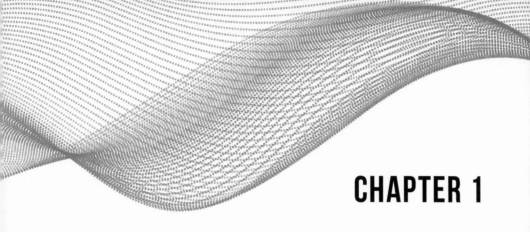

CHAPTER 1

Three Hundred Years of Enlightenment Thinking

Steven Pinker, in his book *Enlightenment Now*, charts the dramatic improvements made in almost every sphere of human affairs since the early part of the eighteenth century. This period, known as the Enlightenment, was the period in which thinkers internalized the ideas which had emerged from the Scientific Revolution of the previous century and broadened the application of reason upon which it was built to all other aspects of human experience and aspiration. I have drawn on Steven Pinker's work extensively in what follows.

The beliefs which controlled the daily lives of people in the Middle Ages are set out by the sociologist Robert Scott:

> Rainstorms, thunder, lightning, wind gusts, solar or lunar eclipses, cold snaps, heat waves, dry spells, and earthquakes alike were considered signs and signals of God's displeasure. As a result, the 'hobgoblins' of fear inhabited every realm of life.

And the historian David Wootton describes the understandings of even an educated Englishman on the eve of the Scientific Revolution of around 1600:

> He believes witches can summon up storms that sink ships at sea ....

He believes in werewolves, although there happen not to be any in England—he knows they are to be found in Belgium .... He believes Circe really did turn Odysseus's crew into pigs. He believes mice are spontaneously generated in piles of straw. He believes in contemporary magicians ....

He believes that a murdered body will bleed in the presence of the murderer. He believes that there is an ointment which, if rubbed on a dagger which has caused a wound, will cure the wound. He believes that the shape, colour and texture of a plant can be a clue as to how it will work as a medicine, because God designed nature to be interpreted by mankind. He believes that it is possible to turn base metal into gold, although he doubts that anyone knows how to do it. He believes that nature abhors a vacuum. He believes the rainbow is a sign from God and that comets portend evil. He believes that dreams predict the future if we know how to interpret them. He believes, of course, that the earth stands still and the sun and stars turn around the earth once every twenty-four hours.

All these beliefs came under the sceptical, rational, and empiricist eye of the Enlightenment thinkers, and today we see the results of this thinking across a huge variety of fields.

Mortality and Life Expectancy

- Child mortality has been dramatically reduced across the world, in Europe from 35 per cent in 1750 to less than 1 per cent today.
- Maternal mortality has been reduced from 1.2 per cent in 1750 to 0.004 per cent today.
- Life expectancy overall has increased from around thirty-five years in 1750 to eighty-plus years today.

Medicine

Over five billion lives have been saved to date, including:

- 177 million through the chlorination of water (discovered by Abel Wolman 1892–1989 and Linn Enslow 1891–1957)
- 131 million through a smallpox eradication strategy (devised by William Foege 1936–)
- 129 million through eight vaccines (discovered by Maurice Hilleman 1919–2005)
- 120 million through the measles vaccine (discovered by John Enders 1897–1985)
- 82 million through the drug penicillin (discovered by Howard Florey 1898–1968)
- 60 million through diphtheria and tetanus vaccines (discovered by Gaston Ramon 1886–1963)
- 54 million through oral rehydration therapy (discovered by David Nalin 1941–)
- 42 million through diphtheria and tetanus antitoxins (Paul Ehrlich 1854–1915)
- 15 million through angioplasty (Andreas Gruentzig 1939–1985)
- 14 million through whooping cough vaccine (discovered by Grace Eldering 1900–1988 and Pearl Kendrick 1890–1980)

It is true that at the time of writing, the COVID-19 pandemic has resulted in some three to five million deaths around the world, but it is also true that within weeks the disease had been identified, and within months successful vaccines had been produced in laboratories in every continent of the world.

Yes, humanity is constantly at the mercy of diseases which can take an enormous toll on life and being, but through the pursuit of medical science, most can be brought under control and many eradicated altogether.

Safety at Work and Standard of Living

It is not only in the sphere of medicine that the effects of following the path of reason and empathy have been felt. In 1929 when the population of the USA was two-fifths the size of today, there were twenty thousand occupational accidental deaths, compared to five thousand in 2015. And in terms of the time required to work 'to make ends meet', in 1929 Americans spent more than 60 per cent of their disposable income on necessities, but by 2016 that had dropped to around 30 per cent.

In overall terms, Steven Pinker comments: 'The combination of a shorter workweek, more paid time off, and a longer retirement means that the fraction of a person's life that is taken up by work has fallen by a quarter since 1960.' This means, of course, that more time is now available for human beings to pursue elements of their human existence which are of particular interest and contribute greatly to an enriched life.

Intelligence

But perhaps of even greater significance is the dramatic increase in intelligence capabilities since the Enlightenment period. Figures from various research models – with the latest conducted by Pietschnig and Voracek involving the meta-analysis of 271 samples from thirty-one countries covering four million people – show that the intelligence, in the broad sense of analytical thinking, of the average person in 1910 would be considered very low by today's standards. This staggering change, though important in itself because of its huge contribution to human flourishing, is particularly important for its contribution to developing and honing a moral compass. It encourages and promotes a much higher level of reflection than could ever have been achieved by our forebears.

Of course, there are still many things yet to be understood about our world and our humanity, but it is undeniable that the rewards of following the path of the Enlightenment have dramatically increased our intelligence, transformed human existence, and contributed hugely to an enrichment of life.

Belief in Supernatural Powers

In the world prior to the Enlightenment, belief in supernatural powers was universal. However, in a 2012 WIN/Gallup poll, 36 per cent of the world's population defined themselves as nonreligious. And in the UK, this has risen to 52 per cent, with only 1 per cent of 18-24-year-olds identifying as Church of England (British Social Attitudes survey for 2018).

Most human beings, at least in the Western Hemisphere, have now left behind any belief in supernatural powers. Our world is simply one planet within a galaxy of galaxies of planets, and we happen to have emerged on it along with trillions upon trillions of other living organisms, each surviving because of how we each manage to harmonize with our particular environment. The intervention of a supernatural power or powers is no longer expected or considered plausible.

In any case, most religions have, as Karen Armstrong points out in her book *A History of God*, continually modified their understanding of 'God' in keeping with their understanding of the world around them. Today, for example, there are few religions in the West which would mount a public defence of the 'fiery furnace of hell' as the fitting punishment for those who have transgressed the will of God; or support the wholesale slaughter by the victor in the name of God of the innocent men, women, and children of the vanquished; or live by the belief that pandemics, floods, droughts, earthquakes, or volcanic eruptions are the actions of an angry God.

In this vein, formal religious leaders, particularly in the West, have quietly modified their beliefs in response to the Enlightenment. But they have done so grudgingly, fearing that changing too much too quickly might bring the whole house tumbling down, plunging us into chaos. For they believe that without God there can be no morality. And without morality, society would collapse.

Origin and Development of Morality

This is possibly the most sensitive area of human experience. Belief in a supernatural God or gods, is precisely that, a belief, whereas morality is an undeniable experience. We experience morality through the actions of others and, by implication, within ourselves. It impacts upon the behaviour of every human being in one way or another. But what is its origin? What light does the Enlightenment throw upon it? Dr. Oliver Scott Curry, research director for Kindlab and a research affiliate at the School of Anthropology and Museum Ethnography at the University of Oxford, a leading authority on the nature, content, and structure of human morality, said, in his Humanists UK Darwin Day Lecture 2021, 'We, as the human species, have evolved to develop a collection of cooperative rules which help us get along together, and promote the common good.' In other words, morality *evolves*: it is part of the same evolutionary process as every other aspect of our being and is in line with the primary goal of the need to survive—indeed, it promotes survival. But it does more; it promotes a profound enrichment of the experience of life itself.

CHAPTER 2

Morality and the Enlightenment

By following the modes of thought released by the Enlightenment, we have made great strides in understanding how morality develops within us. It is a hugely complex story. It involves almost every strand of our being. It draws on studies in many fields – including anatomy, neuroscience, psychology, anthropology, and philosophy. What I offer here is my own layman's understanding of some major contemporary thinking on the subject. As I stated at the beginning of this book, I am not a scholar in any of these fields, but I am someone who has studied and practiced in three professions, engineering, divinity, and education, and who requires, as a necessity for my sense of well-being, a working understanding informed by up-to-date thinking of the rich, complex domain of human nature. What follows may not be pure enough or precise enough for the concentrated scholar, but it has provided me with a meaningful, workable understanding.

Current research on morality, in the primary, broad sense of concern for others, suggests that it arises naturally within us at the dawn of our self-consciousness and, in concert with our innate will to live, is what helps us survive and flourish as a species. It is formed by the myriad of micro-decisions made in the constantly developing human brain. These micro-decisions build huge banks of understandings, which promote our actions and responses in every moment of life. Eventually, a recognizable identity to our being begins to emerge, indicating to the 'other', that is, any other conscious entity, that we are basically loving, fair, trustworthy, and hopeful – or, conversely, uncaring, vengeful, deceitful, and despairing.

A Pictorial Summary of the Making of Morality

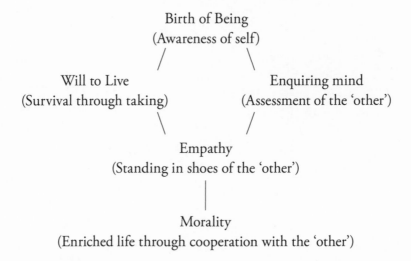

Birth of Being
(Awareness of self)

Will to Live
(Survival through taking)

Enquiring mind
(Assessment of the 'other')

Empathy
(Standing in shoes of the 'other')

Morality
(Enriched life through cooperation with the 'other')

The diagram above attempts to trace the way post-Enlightenment thinking considers the essential elements of human existence, which have evolved over millions of years, to interact, combine and influence one another, to establish, at any one moment in time, a code of behaviour which serves to preserve and enrich the human experience.

At some time around the moment of birth, probably co-incident with the dawn of self-consciousness, two critical elements of our being begin to motivate our behaviour: the simple material, physiological need to survive; and the more complex psychological or spiritual need to satisfy a perpetually enquiring mind.

The need to survive is an a priori need, for without survival nothing has any significance. Therefore, we will do whatever it takes to satisfy this need. Of course, in this task we are in competition with every 'other'. So, the winner will be the strongest and fittest. At this point, the world is a world ruled by physical power in one form or another.

However, the enquiring mind is very curious about this world of the 'other', the world outside self, and cannot resist asking the question 'why?' about the behaviour of the 'other'. Gradually, this questioning begins to build a picture of the 'other' and allows the questioner to stand in the shoes of the 'other' and, to some degree, experience what the 'other' experiences. In other words, alongside the experience of consciousness of personal being there is an experience, at least in part, of the consciousness of the 'other'. Through this empathy, the recognition grows that rather than allow the exercise of power, force, or aggression to determine an issue, with the inevitable outcome of defeat for one party accompanied by pain, unhappiness, rejection, and possibly death, much more can be accomplished, to the good of the well-being of each party, by cooperation.

Hence, a moral code develops encapsulating the most acceptable behaviour to bring about, not only survival, but a great enrichment of life for all concerned.

The development and nourishment of our own moral identity is therefore essential if we are to take a secure, meaningful place in the human community.

What I am proposing in this book is that we need, each of us, to pay great attention to the development of our own moral compass housed deep within our beings, for it is only with such a compass that we will steer positively and successfully through the complexities of human life. And further, that our compass must be regularly fine-tuned throughout our lives by our continuous reflection upon the consequences of our every action and the actions of others.

PART ·TWO

MORALITY TODAY

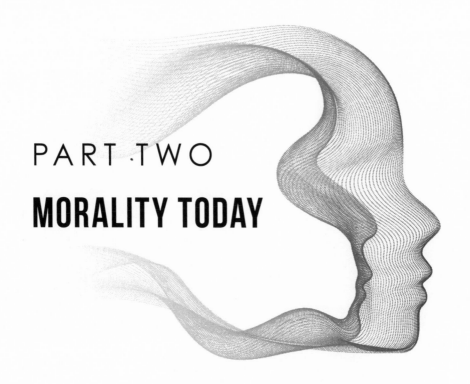

Without civic morality communities perish, without
personal morality their survival has no value.
Bertrand Russell, *Authority and the Individual*

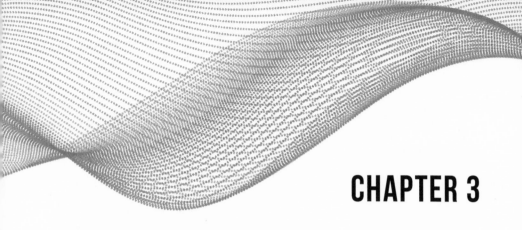

CHAPTER 3

The Existential Necessity for a Moral Compass

I have a miniature bust of Socrates on my desk which I brought back from a visit to Turkey. I look at it often, because it was a quotation of his, which I came across in my late thirties, that saved my life. Not my physical, biological life (although, who knows, maybe that too!), but the life of my inner being. From my youth I have had an overwhelming (some would say, overbearing!) need to understand and explain. Anything and everything, no matter the time or place, was subject to this enquiry – or onslaught. It won me few friends, and it troubled me.

Until, that is, I came across these words of Socrates: 'An unexamined life is not worth living.' For me, this was the encouragement I needed. Not that I am a thinker in the class of Socrates; far from it. But those words of his helped me to accept the 'me' that I am, the *me-that-only-I-am-aware-of*, and perhaps, through the application of those words of his to *that* 'me', to become a more responsive 'me', more accepting, less assertive, more at ease with the world.

The Soul as the Home of 'Being'

Modern research suggests that as the human brain develops its immense information-processing capacity, the accumulation of the millions of decisions it makes day by day eventually presents the individual as a single, autonomous, recognizable entity to other human beings. The defining feature of this entity is what is commonly referred to today as *mindset,* or *character,* or what has been referred to in the past, within a universal belief in God or gods, as *soul.* Either way, mindset, character, or soul, represents the essence of a human being, or the *me-that-only-I-am-aware-of.*

As I explained earlier, I prefer the word *soul* – though in a purely human dimension without any supernatural connotations – as a more rounded, warmer, comprehensive embodiment of the amazing entity of the human presence. Current research suggests that this entity that I call soul is the living, dynamic, moral essence of who I am, created through the continuous exercise of the will at countless crossroads encountered in the brain, in response to the continuous bombardment of experiences and actions in the material world.

The ability to recognize and rely upon a predictable, overall moral response from other human beings creates not only a secure environment for me but also a strong, stable, healthy, satisfying human society. It is this moral dimension of human existence that provides the necessary environment for the full flowering and flourishing of the human species.

The Crucial Role of Empathy in the Making of Morality

An essential characteristic of being for the development of morality is empathy. This critical characteristic of human awareness seems to begin to develop at a very early age and is evident to varying degrees across all cultures.

A study by Professor Rebecca Saxe and Ashley Thomas, of the Massachusetts Institute of Technology in the US, published in the journal *Science* in January 2022, found that babies, across cultures, use the intimate sharing of saliva – through kisses and drooling – to determine those who will respond urgently to their cries of distress. This simple evaluation mechanism is crucially important as their very survival depends on such attention.

Further, Oliver Scott Curry, Research Director for Kindlab, and Research Affiliate at the School of Anthropology and Museum Ethnography, University of Oxford, tells us that modern research shows that a degree of empathy is clearly discernible in children as young as four months old.

Also, Zahn-Wexler et al, in an article 'Development of Concern for Others' in 1992, state that children as young as a year and a half spontaneously give toys, proffer help, and try to comfort adults or other children who are visibly distressed. It's as if they identify themselves in the 'other'. This is the beginning of empathy, without which no moral development is possible.

And modern research has found that some concept of morality is evident throughout the human species. D. E. Brown concludes in his book *Human Universals* (McGraw-Hill, 1991), that people in all cultures distinguish right from wrong; have a sense of fairness; help one another; impose rights and obligations; believe that wrongs should be redressed; and proscribe rape, murder, and certain kinds of violence. Oliver Scott Curry also concludes from his research: 'We, as the human species, have evolved to develop a collection of cooperative rules which help us get along together and promote the common good' (Darwin Day Lecture 2021, Humanists UK).

Modern research therefore suggests that 'empathy' is an openness of 'being' to the feelings and sensations of the 'other'; is an essential prerequisite for morality; emerges very early in human development and is evident across all cultures.

Don't we all recognize the roles of empathy and morality in our everyday lives when we describe as a 'good person' the person who has an affection and respect for all life; who lives modestly; who always has a concern for others; who always seeks to understand and is compassionate in justice? Are not these the people we respect, the people we are drawn to?

'Depth of Being' Is at the Heart of All 'Religious' Experience

I am convinced that it is that ultimate 'depth of being', the *me-that-only-I-am-aware-of,* which is at the heart of all religious, that is to say spiritual, experience. As I attempted to show in my last book *Faith Refractioned*, it is my belief that religion begins for humankind at the dawning of consciousness and the simultaneous realization that there is something outside the self which must be explored.

As we grow, we gradually realize that this awareness of the 'other' includes not only the physical, material world surrounding us, but also the non-material, metaphysical world. We become aware not only of the awe arising within us as we gaze at the stunning beauty of the natural, 'outer' world – from the mountaintop, or the lakeside, or at the breaking of a new dawn, or at the sunset of another day – but also of the awe inspired by 'inner' experiences such as an exquisite musical composition, or an act of supreme self-giving, or a moment of sublime imagination.

All these experiences of the 'other', in whichever dimension, touch the inner core of our beings where, through innumerable interactions with a myriad of pulsating elements of primal energy, the *me-that-only-I-am-aware-of* is born and refined, and I become a unique living being, a being with a distinguishable moral identity, yet at the same time, a being which is congenitally connected to the living heartbeat of the world around me.

My True Self, My 'Soul' – an Organic Relationship with the Elemental Universe

As I have suggested above, the single word which, for me, expresses this sensation of ultimate, unique 'me', independent though at the same time congenitally connected to the world around me, is *soul*. This 'encounter within being', at the deepest of all possible levels of self, is an encounter with the ultimate me, the 'soul', and it is the most profound determinant of the nature and quality of our lives. It is our soul, formed through millions of interactions in the depths of our beings in response to the impinging presence of 'otherness' in which we are immersed, which determines the things we value, the choices we make, the security and peace we feel, the happiness we experience – in short, all the things which our fathers before us associated with our being 'children of God'. But it is not an encounter with a 'being without' but an 'encounter within', an 'organic relationship' of the most profound kind, between the *me-that-only-I-am -aware-of* and the elemental essence of the universe around me. This organic relationship is my true being, it is who 'I' fundamentally am. And it is this organic relationship which we, as human beings, must learn to explore and grow and refine if we are to experience contentment and joy in our existence. Otherwise, we will be constantly 'tossed to and fro and carried about with every wind of doctrine' as St Paul puts it, and we will not be 'at ease' in our existence: we will only know 'dis-ease'.

Nurturing the 'Soul'

How do we develop or cultivate this 'soul' or 'inward domain of consciousness', as the philosopher John Stuart Mill called it? If it is in our heart of hearts, as we commonly say, that we pass our final judgements on happenings and experiences; where we decide what is annoying, disgusting, scary, hurtful, offensive, evil and also what is attractive, beautiful, heart-warming, loving, good – in essence, where we make the ultimate choice between good and evil – then we must take every care to nurture and bring to maturity this most crucial kernel of our beings, the *me-that-only-I-am-aware-of*, and its most intimate relationship with the world in which it is immersed.

From our birth we are quick to learn that the physical aspects of our being require constant sustenance and attention if we are to survive at all. But, because the consequences of inattention to the 'being within' are not so dramatically immediate or obvious – as they are to any inattention to our physical, 'outer being' – we make little or no attempt to nurture its development. Yet the ultimate consequences of this inattention are of profound, life-long significance. It is the difference between a life merely lived and a life lived in all its fulness.

To withstand the daily rigours of human existence as independent, wholesome human beings, rather than shapeless blobs of haphazard, uncontrolled reactions, we must build our lives on deep, solid foundations, rather than on shifting sands: in our heart of hearts, in our souls, we must strive to become wise and strong, discerning in our judgements and generous in our compassion. To do this, we must explore and nurture and refine the 'inward domain of consciousness' by serious reflection, not only on our own experiences in all their dimensions, but also on the experiences of others who have travelled the same road and made the same discovery.

The First Step – Choose to Love

The first and most important step that we must take, which acts as the catalyst for the whole life-changing process or, to use a familiar metaphor, which acts as the leaven in the dough of the bread of elemental existence, is to make the decision to *love the 'other'*: to adopt, as I spell out in my previous book *Faith Refractioned*, a deep inner attitude of being which 'sees all life through the eyes of love and lives all life through the power of love'. This alone will enable the *me-that-only-I-am-aware-of*, my soul, to experience itself in harmony with the heartbeat of the world in which it is immersed. I must not see my existence as a continuous battle against the elemental powers surrounding me, but rather as a positive engagement leading to a harmonious relationship to be enjoyed.

Constant Action-Reflection

Feeling this heartbeat will illicit actions, and these actions will arouse reflection, and from these reflections, changes to our future responses will be made. This process of action-reflection, consequent upon our decision to love, is what builds and refines the soul. The process follows the same steps as in the perfecting of any skill in any sphere of life: after each attempt, following rigorous analysis and reflection, adjustments are made. In every sphere, following this process, a huge enrichment of being is experienced – 'life in all its fulness', as Jesus of Nazareth would say.

This constant process of action-reflection is not easy, it demands much effort – as do all efforts to achieve improvements and perfection in any endeavour. Digging down to the secret places of our hearts, to the domain of the *me-that-only-I-am-aware-of*, is not for the faint-hearted, nor the less-seriously minded; but if we wish to change our lives, to experience them and live them in a different way, it is the place where change will need to occur. Because it is in our 'heart of hearts', our souls, that we pass judgements on happenings and experiences in our day-to-day existence; it is in the depths of our beings that we decide, as I have said previously, what is annoying, offensive, disgusting, hurtful, scary, chilling, evil; and it is in the depths of our beings that we decide what is attractive, beautiful, heart-warming, loving and good. It is the *me-that-only-I-am-aware-of* that makes those evaluations and experiences the outcomes.

The Moral Choice

In every situation in life, we have a choice: to do the *easy* thing, or to do the *right* thing. Sometimes – but not often – doing the easy thing is also doing the right thing. But, as Jesus of Nazareth said: 'The gate is wide for those doing the easy thing and narrow for those doing the right thing – and few will go that way.' But every day, that is the choice we have to make.

The most important question we need to address therefore is: how do we nurture and refine this most crucial part of our beings, the *me-that-only-I-am-aware-of,* the soul? How do we promote its growth and bring it to its maturity? How do we cultivate the beautiful art of action-reflection, delicately creating the emotions deep within our hearts, that ultimately become our very being?

The Need Is Urgent

The need could not be more urgent! Humanity is entering a new era: the era of artificial intelligence (AI), where computers will infiltrate and be directly commissioned to control and direct almost every aspect of our lives. Once a code has been embedded into a device, it will complete its task without any further human intervention.

Even today, many of the tasks necessary for the efficient running of modern civilizations have already been codified, bringing undoubted benefits to huge swathes of human societies – such as the elimination of dangerous, unhealthy, tedious work practices; the almost instantaneous scrutiny of millions of possibilities in chemical sequencing to facilitate the control or elimination of many hitherto unconquerable diseases; immediate access, for almost every human being, to virtually limitless information; and instant communication with people anywhere in the world.

Yet we are only at the beginning. The next step, to full AI, is probably going to be the most revolutionary. And since change in this new world of digital technology is exponential rather than linear – as it was for thousands of years in the 'old' world – then we have little time to address the issues I have raised above. The fields that are now in the sights of digital technology will provide opportunities, at many points deeply embedded within the programs, for the expression of the value judgements and moral dispositions of the designers.

Take for example the self-driving motor car: suppose the manufacturers of such cars were personally so strongly opposed to, say, factory farming or the human consumption of any animal products, that they programmed their cars to refuse to take passengers to any fast-food drive-through restaurant? Or consider the self-driving vehicle, which, because of a value-driven cost benefit analysis buried deep within its program, elects to run down a child, regardless of the certain fatal consequence, because it assesses that the 'soft' obstacle of the child would cause less damage to the car

19

and occupants and therefore less cost to repair, than swinging aside into a 'hard' bollard? Or the 'surgeon' robot, which 'refuses' to carry out legal abortions because its designer's religious beliefs against abortions have been subtly placed deep within its program?

These are but a few simple examples of where belief systems or values, lying deep within our inner beings, could influence, either consciously or unconsciously, the performance of the products we make.

The technology lying just around the corner will provide opportunities and threats on a massive scale, and will necessitate, to a historically unsurpassed degree, not only extremely robust legislative control, but also a strong, universally acclaimed moral code anchored deep within the souls of the designers and manufacturers, if immoral conduct and deviancies are not to cause havoc on an unimaginable scale.

As Plato wrote in *The Republic*, having thought deeply about the nature of human beings and human societies: 'There will be no end to the troubles of states ... till philosophers become kings in this world, or till those we now call kings and rulers really and truly become philosophers.' And the 'kings and rulers' of the incredible world just ahead of us will be not only the political rulers but also the digital designers and entrepreneurs. It is of critical importance, therefore, that these new world kings and rulers are people of virtue with strong moral values, if the world is not to slip into chaos and darkness.

But it is not just a matter of the character of *their* beings, but also of *our own* being, every one of us as citizens, because it will be up to us, as citizens of the world, to exercise the constant vigilance which alone will protect our integrity and freedoms. It must not be left in the hands of designers and rulers to decide.

Hence the critical importance of devoting significant energy to the development, in each one of us, of our moral sensibility, of our own personal moral compass. Whenever a decision is to be made or an attitude adopted, our moral compass will determine whether it is for good or evil, whether it will bring hope or despair, whether it will be constructive or destructive, whether it will make a positive contribution to the enrichment of our lives or lead to the despoilment of our great potential.

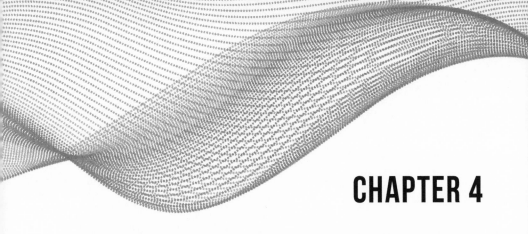

CHAPTER 4

Characteristics of a Moral Compass

There are many studies, now, which suggest the universality of the personality traits or characteristics of a human being. It is true that the significance of any characteristic may vary depending on several factors, not least of which is the level of economic development of the community to which they belong. As the psychologist A. H. Maslow points out, there is little incentive to satisfy the 'higher needs' for social interaction, personal status, or self-actualization until the immediate physiological 'basic needs' for food, clothing, shelter, and safety have been met.

Nevertheless, few in the world today, if asked for the characteristics of a 'good' person, would not suggest someone who was *loving, fair, trustworthy, hopeful* and, conversely, would not suggest the characteristics of a 'bad' person to be someone who was *uncaring, vengeful, deceitful,* and *despairing.* A simple way to confirm the validity of these characteristics is to ask yourself the question: 'What sort of person would I like my son or daughter to choose as a partner in life?'

And if such moral characteristics are important to us as individuals, they are also crucial for us as societies. Without a moral compass we will revert to the default position of 'survival of the fittest', which means, for us as individuals, achieving our ends by any means necessary, including force and violence; and for us as societies, ultimately meaning war.

Each of these overarching characteristics contains sub-traits to varying degrees of refinement as set out below: -

Loving – compassionate, self-less, forgiving.

Fair – understanding, un-biased.

Trustworthy – honest, reliable, dependable.

Hopeful – confident, joyful, happy.

Uncaring – selfish, lustful.

Vengeful – spiteful, violent.

Deceitful – cheating, lying.

Despairing – self-loathing, depressed.

I have attempted to capture these traits and sub-traits diagrammatically in the form of a moral compass. The neutral position is with the hand pointing upwards, for this is where we begin life's journey. Our moral character will be shaped by the predominant position taken up by the hand in dealing with the multitude of circumstances and issues encountered along the way – though occasionally, the hand may move in a different direction should we act 'out of character'. The more we reflect seriously about each circumstance and issue, the more will the hand traverse the right-hand side of the compass.

A Pictorial Presentation of a Moral Compass

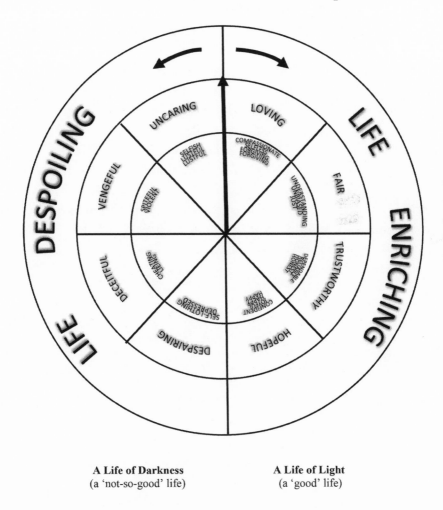

A Life of Darkness
(a 'not-so-good' life)

A Life of Light
(a 'good' life)

The moral compass needs to be embedded deep within the very heart of the *me-that-only-I-am-aware-of.* And it needs to be constantly refreshed and refined. This nurturing is achieved through the dialectical process of action-reflection. Action alone is not enough; each action needs to be directed and refined through reflection. As this process is repeated, hour after hour, in situation after situation, the 'natural' alignment of the compass begins to take shape deep within the soul. As its direction is repeatedly set, it eventually gives a recognizable identity to the whole being. Such a refined compass is essential if we are to successfully navigate the ever-changing, unpredictable sea of life: we need it not only to survive but to live well, to live meaningfully, to have a purpose and shape to our existence, and to feel a peace within our souls.

Thankfully, we are not alone in this task. The sages, historians, and writers through the ages, together with the accumulated practical wisdom of tribes and communities, provide us with a rich resource. All that is required of us is serious, continuous reflection, and a determination to live beyond mere survival, to a fulfilling, contented life.

Reflecting upon the issues, incidents, stories, and proverbs set out below, themselves only a brief selection from all that is available throughout the world, does, I believe, nurture, and grow the *me-that -only-I-am-aware-of,* the soul. Constant refinement will gradually bring the soul to a level of maturity fit for purpose, not only in the world of today, but also in the world of tomorrow – the incredible, challenging new world which lies just around the corner.

PART THREE

DEVELOPING AND NURTURING OUR MORAL IDENTITY

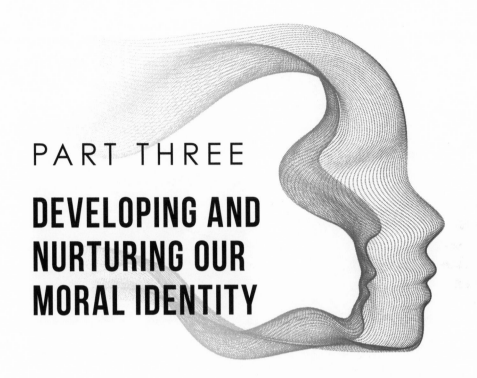

No great improvements in the lot of mankind are possible, until a great change takes place in the fundamental constitution of their modes of thought.
John Stuart Mill

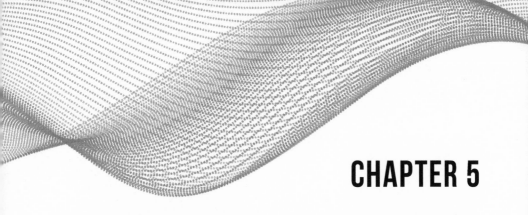

CHAPTER 5

The Need to Take Time Out to Reflect

What follows are some of the stories, poems and sayings that have caused me to stop and reflect over the years.

It is easy to be swept along by the hurly-burly of life – passing exams, finding a job, getting married, building a home, rearing children, without ever taking time out to consider if any of it has meaning, or if any of it matters, or if it matters, why it matters.

We rarely stop to consider the words of the Welsh poet **W. H. Davies**, in his poem 'Leisure', written in 1911, at the full height of the Industrial Revolution:

> What is this life, if full of care
> We have no time to stand and stare.
>
> No time to stand beneath the boughs
> And stare as long as sheep or cows.
>
> No time to see when woods we pass
> Where squirrels hide their nuts in grass.
>
> No time to see in broad daylight
> Streams full of stars, like skies at night.
>
> No time to turn at Beauty's glance,
> And watch her feet, how they can dance.

No time to wait till his smile can
Enrich that smile her eyes began.

A poor life this if, full of care,
We have no time to stand and stare.

Or take time out to consider the words of **St Paul** in the closing sentences of his letter to the fledgling Christian community in the town of Philippi, a Roman colony, in Macedonia, (today, northern Greece):

> Summing it all up, friends, I'd say you'll do best by filling your minds and meditating on things true, noble, reputable, authentic, compelling, gracious - the best, not the worst; the beautiful, not the ugly; things to praise, not things to curse.
>
> (Philippians 4.8-9)

Or give a thought to these words from the Book of Proverbs commonly attributed to **King Solomon** and compiled about 500 BC:

> Listen, friends, to some fatherly advice;
> Sit up and take notice so you'll know how to live.
> I'm giving you good counsel;
>   don't let it go in one ear and out the other.
>
> When I was a boy at my father's knee,
>   the pride and joy of my mother,
> he would sit me down and drill me:
>   'Take this to heart. Do what I tell you – live!
> Sell everything and buy Wisdom!
>   Forage for Understanding!
> Don't forget one word!
>   Don't deviate an inch!
> Never walk away from Wisdom – she guards your life:
>   Love her – she keeps her eye on you.
> Above all and before all, do this: Get Wisdom!
>   Write this at the top of your list: Get Understanding!

Throw your arms around her – believe me, you won't regret it;
  Never let her go – she'll make your life glorious.
She'll garland your life with grace,
  She'll festoon your days with beauty.' (Proverbs 4.1-9)

Or reflect on the words of **Steven Pinker** in his book *Enlightenment Now* (2018):

> People can be healthy, solvent, and literate and still not
> lead rich and meaningful lives … (we) worry that all that
> healthy lifespan and income may not have increased human
> flourishing after all, if they just consign people to a rat
> race of frenzied careerism, hollow consumption, mindless
> entertainment, and soul-deadening anomie. (p. 247)

To avoid this, Pinker suggests that we use 'information to resist the rot of entropy and the burdens of evolution', and that 'The supernova of knowledge continuously redefines what it means to be human. Our understanding of who we are, where we came from, how the world works, and what matters in life depends on partaking of the vast and ever-expanding store of knowledge' (p. 233).

As human beings then, it is essential, in every age, to take time out to reflect, if we are to achieve our overall goal of maximizing human flourishing. Reflecting on incidents, issues, dilemmas, and developments in any sphere will help us continuously develop and hone our moral sensibility. In this way we shall be constantly reviewing, refining, and stabilizing our moral compass.

And the place to start is to make the choice to open our souls, express our empathy, and *follow the way of love* through all the vicissitudes of life. Yes, it is a choice – despite the difficulties created by the environment around us. It is true that for some the environment will be crushing, whilst for others it will be light. But it is still always a choice, either to live selfishly, with little regard for others, or to live caringly, always striving to take others into account.

This decision forms the first segment of the right half of the moral compass – for without love deep within our beings, there can be no light, no flourishing, at all.

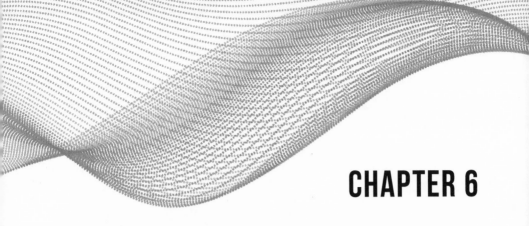

CHAPTER 6

Outstanding Lives, Stories, Poems, Sayings, and Songs

All human beings live within the language and culture of their own time. Of course, some of their language and culture will have been inherited from former generations, with each of those generations in turn developing their own unique contributions in line with their changing knowledge and experiences. Therefore, the apprehension and understanding of life in any age will, of necessity, be circumscribed by the thought-forms and language of that age. For example, the stories, writings, and experiences of my great-grandparents are devoid of any words or thought-forms derived from the world of air travel, television, computers, or internet. Yet today we can hardly utter one sentence without reference to one or another of these realities.

And since we rely heavily on worldly metaphors to express our experience and understanding of abstract matters such as religion, morality, relationships, motivating drives – all matters of the 'heart' – it is not surprising that the thought-forms and expressions used by our forebears may not be immediately appealing to us today. This does not mean, however, that their *experience* of such matters was not authentic. It simply means that we will be couching *our* understandings of these *same* experiences in language and thought-forms current in our own day. Hopefully this will also mean that our understandings of these experiences will be enhanced.

All this must be borne in mind when we turn to some of the great historical literary works to help us nurture and grow the moral compass of our souls. Frequently our predecessors will attribute to 'the voice of God' some inner feeling that a particular action should be taken, or a change of direction adopted, or an admonition for wrongdoing accepted. But the words and concepts, though foreign now to our way of thinking, do not invalidate the sentiments experienced. It is by exposing our own beings to such experiences, either through experiencing them for ourselves or learning from the experiences of others, that the moral compass of our souls is formed. Usually, the change wrought by this process is gradual and subtle. On occasions, however, it can be quick and dramatic as in St Paul's 'conversion' on the road to Damascus.

Sometimes, as we reflect upon the world, it does seem that the human species is stumbling around in the dark, grovelling in the mud and grime, continually provoking turmoil, confusion, bloodshed, and fear. And yet, also, sometimes, even amid all that, we are coming to recognize, through some particular experience – perhaps an act of supreme courage or a refusal to accept the culture of darkness or a sacrifice made by one of our number – that we are all capable of *higher* things. These evidences tell us that as human beings we have the choice to live either in the 'darkness' or in the 'light': we can see the world as essentially a place of evil and destruction, or a place of goodness and hope.

It is a simple fact that those who love are always positive about the world and life. And feeling positive and hopeful about the world is the springboard for reaching out to others to help them find – in some small measure – an enrichment of *their* lives.

For it is also true that we don't move from living in darkness to living in light simply because we *say* the words. We must *live* the words; we must *live love*. We must allow it to permeate deep into our souls, such that we naturally see all life through the eyes of love and live all life through the power of love.

If you have experienced genuine love, you will know that it turns your life upside down! It affects how you see the world, how you respond to it, how you feel about it. It literally changes *you* – and everyone around you can see it! Now that's a remarkable thing because the world around you hasn't changed at all. It's simply the effect of a change in your heart, of becoming 'open' to the 'other'.

The following stories, poems, and songs have helped to shape the moral compasses of generations of people across the world. By seriously reflecting upon them, we can shape and hone *our* moral compasses too and so help us all flourish as individuals and societies.

1

Love Changes Everything

Andrew Lloyd Webber, in his musical *Aspects of Love* has the young lover Alex sing a song which has this very title:

> Love, love changes everything
> Hands and faces, earth, and sky.
> Love, love changes everything,
> How you live and how you die.
> Love can make the summer fly,
> Or a night seem like a lifetime,
> Yes love, love changes everything,
> Now I tremble at your name, for …
> Nothing in the world will ever be the same.
> Love, love changes everything
> Days are longer, words mean more
> Love, love changes everything
> Pain is deeper than before.
> Love will turn your world around
> And that world will last forever
> Yes love, love changes everything
> Brings you glory, brings you shame
> Nothing in the world will ever be the same.
>
> Off into the world we go
> Planning futures, shaping years
> Love bursts in and suddenly all our wisdom disappears.
> Love makes fools of everyone
> All the rules we made are broken
> Yes love, love changes everyone
> Live or perish in its flame
> Love will never ever let you be the same.

2
Tribute to the Poet John Keats

In 2018 **Sir Bob Geldof** wrote this tribute to the poet John Keats:

Whatever alchemy of articulacy Keats has created, there are few who seem to understand the incoherent agony of soul-loss or soul-abandon quite as he does. Who seem to know what love is. What it looks, feels, tastes and smells like. And who can magnificently, miraculously, gloriously transliterate emotion and its sense into numinous transcendent words.

In truth it is only love that makes transcendent and essential the otherwise pedestrian condition of the human being … Keats's aim was to awaken in people their higher spiritual nature … For Keats, this could only be achieved through love, which for him, was the essence of man and sacred.

3
First Letter of St John in the New Testament

My dear friends, I'm not writing anything new here. This is the oldest commandment in the book, and you've known it from day one. It's always been implicit in the Message you've heard. On the other hand, perhaps it is new, freshly minted as it is in both Christ and you – the darkness on its way out and the True Light already blazing!

Anyone who claims to live in God's light and hates a brother or sister is still in the dark. It's the person who loves brother and sister who dwells in God's light and doesn't block the light from others. But whoever hates is still in the dark, stumbles around in the dark, doesn't know which end is up, blinded by the darkness.

(1John 2.7-11)

4
Selection of Biblical Passages concerning Living in the Light

Give me your lantern and compass,
    give me a map,
So, I can find my way to the sacred mountain,
    To the place of your presence. (Psalm 43.3)

This is the kind of fast-day I'm after:
    To break the chains of injustice,
    Get rid of exploitation in the workplace,
    Free the oppressed,
    Cancel debts.
What I'm interested in seeing you do is:
    Sharing your food with the hungry,
    Inviting the homeless poor into your homes,
    Putting clothes on the shivering ill-clad,
    Being available to your own families.
Do this and the lights will turn on,
    and your lives will turn around at once. (Isaiah
    58.6–8)

The Life-Light blazed out of the darkness;
    The darkness couldn't put it out. (John 1.5)

You are the light of the world. (Matthew 5.14)

5
The Small Beatitudes – A Modern Take

Blessed are those who can laugh at themselves;
    they will have no end of fun.
Blessed are those who can tell a mountain from a molehill;
    they will be saved a lot of bother.
Blessed are those who know how to relax without looking
for excuses;
    they are on the way to becoming wise.

Blessed are those who are sane enough not to take themselves too seriously;

> they will be valued most by those about them.

Happy are you if you can take things seriously and face serious things calmly;

> you will go far in life.

Happy are you if you can appreciate a smile and forget a frown;

> you will walk on the sunny side of the street.

Happy are you if you can be kind in understanding the attitudes of others

> even when the signs are unfavourable;
>
> you may be taken for a fool, but this is the price of charity.

Blessed are those who think before acting and pray before thinking;

> they will avoid many blunders.

Happy are you if you know how to hold your tongue and smile,

> even when people interrupt and contradict you or tread on your toes;

the Gospel has begun to seep into your heart.

> Above all, blessed are you who recognize the Lord in all that you meet;

the light of truth shines in your life for you have found true wisdom.

(Joseph Folliet, 1903–1972)

6
The Shepherd Boy and the Giant Warrior

You've probably heard the biblical story of **David and Goliath** – how David, a mere shepherd boy, though very skilled with a stone and sling, saved the king of Israel and his army from the might of the Philistines. It is a very positive story reminding each one of us that no matter how small and insignificant we see ourselves to be, we can, with dedication, love, and courage, achieve great things.

The two armies faced each other on opposite hillsides, and as was the custom at the time, they agreed to try and settle matters first by single combat. Each army would send out their best warrior to fight each other, and the winner would claim total victory of the war. The Philistines were hugely confident, because they had a warrior of immense size and very skilled in combat. He was an utterly fearsome sight, clad from head to toe in metal armour! Not one of the Israelite army was prepared to take up the challenge, and the Philistines mocked them for their cowardice!

And that was the situation when David the shepherd boy arrived with provisions for his brothers who were fighting in the king's army. No one would take up the challenge of fighting the giant Goliath. So, David did! He refused to listen to the ridicule of the soldiers gathered around him and refused the armour he was offered. He was supremely confident in his ability to bring down Goliath – with his stone and sling! After all, he was a successful shepherd boy, and it was with his stone and sling that he constantly defended his sheep. So, he walked boldly out to face Goliath. As the giant Goliath taunted him, David loaded his sling and fired. His stone hit Goliath straight between the eyes, and down he tumbled!

The king and his army cheered and cheered – the king, because it meant he could now take over all the lands of the Philistines, and the soldiers, because they wouldn't now be killed or maimed in battle! It was a great day and was marked in the annals of the history of Israel.

—

Yes, the story of David and Goliath is a great story; it teaches us that we don't have to be a 'Goliath' to have the skills necessary for a successful life. However small our endowments from life, we too can achieve great things, if we learn how to develop our abilities to the full.

7

The Shepherd Boy and the Prince

There is an equally significant story in the Bible, which is perhaps not quite as familiar to you, about the enduring friendship between **David and Jonathan.**

Jonathan was the king's son, David a shepherd boy, yet although they came from entirely opposite levels of society, they struck up an immediate friendship which endured through many awkward, sometimes dangerous, and often turbulent times. Jonathan was the king of Israel's son and heir – except that Saul, the king, had quickly realized after the victory over Goliath that David would be the one to take over his throne after him!

Saul first promoted David to be his 'special assistant' because David possessed a special ability, through his harp-playing and singing, to sooth Saul's soul in his frequent bouts of torment and despair. But David was also a very accomplished soldier and leader of men, so it wasn't long before Saul promoted him again, this time to be commander of his troops. This was a highly popular promotion, enthusiastically applauded by both the men and the public alike. Indeed, when returning from battles, the cheers were louder for David than for the king himself!

Gradually, however, the king became jealous of David's popularity, and the jealousy grew in his heart to erupt in great fits of rage against David, even to the extent of plotting to kill him.

Throughout all this, Jonathan's love of David never faltered. He did everything he could to protect his 'brother' – as he had come to regard him – from the erratic moods of his father, even helping him evade, on numerous occasions, his father's murderous intentions. Although he had every reason to feel aggrieved by the great popularity of David, and the recognition by everyone that, one day, it would be David and not he who would sit on the throne of Israel, Jonathan felt no anger, bitterness, or jealousy toward David. He was simply happy to stand by David's side and support him in any way he could. Within moments of their first meeting, Jonathan had sworn to be David's blood brother and had sealed his love for David by giving him his robe, his sword, his bow, and his belt, the most important and personal of his possessions. It was a symbolic act, pledging to David that he would never use position or force against him. And he kept his word.

—

This is true friendship, this is real love, and it literally turned both their worlds upside down.

8

Mother and Daughter-in-Law

The story of **Naomi and Ruth** is set around three thousand years ago, first in Bethlehem in Canaan and then in Moab, which is on the opposite side of the Red Sea.

There was famine in Canaan which lasted so long that Naomi, her husband, and their two sons took the very big decision to pack up, leave Bethlehem, and go to Moab, where they had heard there was plenty of food. It was quite a wrench leaving Bethlehem because it was the place of their birth and the land of their fathers, a land which they believed had been given to them by their God. Moab was a very different land with different traditions and different gods. But they soon settled down and began to build a new life for themselves.

Then disaster struck. Naomi's husband died, leaving Naomi to bring up their two sons alone, in a foreign land. She struggled on, and eventually her two sons, with her blessing, married two local girls, Orpah and Ruth, and Naomi lived happily with them as they all settled down together.

Then a second disaster struck! Naomi's two sons were killed in an accident at work, leaving Naomi alone again. Except she didn't *feel* alone, for this time she had her two daughters-in-law, whom she loved very much, and who loved her deeply. Still, in the traditions of the day, life was very tough for widows.

Then one day, Naomi heard that the famine in Canaan was over, and Bethlehem was prospering again. Naomi pondered what to do: back in her hometown she owned a small field and had relatives who would look out for her. On the other hand, she knew she was lucky to have two such loving daughters-in-law, and they would naturally wish to stay near their own mothers and relatives. Finally, she decided she must return to Bethlehem, to her own people. But the two daughters-in-law would have none of it; if Naomi was leaving then they would go with her! Naomi was very moved by their devotion and love but insisted they should stay and make new lives for themselves among their own people. Eventually, Orpah agreed, but Ruth, who had a particularly deep love of Naomi, insisted that she go with Naomi, even though it meant giving up everything she had known from birth and trying to forge a new life in a foreign land with different customs and traditions. She said to Naomi: 'I want to go wherever you go and to live wherever you live; your people shall be my people, and your God shall be my God.' (Ruth1.6) So, they both set off for Bethlehem.

Life was very hard for the two women on their own; they barely scraped a living from Naomi's field. So, one day, at the beginning of the harvest, Ruth decided to go to a neighbouring field and pick up the grain not scooped up by the reapers – a practice accepted by all. Naomi was not very happy about this as she knew how vulnerable a beautiful young woman would be in a field on her own. So, she arranged instead for Ruth to go to the fields of her late husband's relative, Boaz, whom she knew to be a good man who would treat her well and protect her.

So, Ruth went to the fields of Boaz and toiled all day without a break. When Boaz saw her working so hard, he arranged with his reapers to leave a little extra grain behind so that Ruth would always return home with a full bag. This was his way of recognizing his kinship and respect for Naomi and Ruth.

Gradually, a deep affection developed between Ruth and Boaz, encouraged by Naomi, who realized what a good thing it would be for Ruth to find love again. For Boaz, however, there was an obstacle to him taking Ruth as his wife: there was another man who had a prior claim! It was their custom that when a male member of a family died, the next closest member of the extended family had the right to take possession of all lands and marry the widow. Unfortunately for Boaz, there was a relative before him who would have the first claim, even though that relative had not yet stepped forward.

So, Boaz, being an honourable man, sought out the first in line and informed him of the situation. The relative quickly agreed, but when Boaz went on to explain the full legal implications – that if he had a son with Ruth, it would be that son and not one of his own sons, whom he already had to another wife, who would inherit everything – the relative changed his mind and gave up his right in favour of Boaz.

So, Boaz married Ruth, and they did indeed have a son – much to the delight of Naomi, who looked after him as if he were her own. And this son became the grandfather of David, Israel's greatest king!

—

What lies at the heart of this story is the moving love which Naomi and Ruth had for each other. They were from different lands with different cultures; they worshipped different gods and were from different strata of society. Yet the love they shared overcame all their differences and turned their lives upside down.

9
Romeo and Juliet

This is **William Shakespeare's** story of two young lovers who came from the two chief families in Verona, the Capulets and the Montagues, whose utter dislike for each other was known throughout Verona. Even a chance meeting in the street of the servants of each family, which frequently happened, always ended in brawling and bloodshed. Unfortunately, Romeo was a Montague and Juliet a Capulet!

Initially, Romeo only had eyes for a young girl called Rosaline, the niece of Lord Capulet. He was besotted with her, but she ignored him and treated him with little curtesy or affection. This badly upset Benvolio, Romeo's cousin and very good friend, for Romeo simply could not see how badly Rosaline was treating him.

One day, Lord Capulet put on a big party to which all the fair ladies and nobles of Verona were invited. And Rosaline was one of them. Benvolio saw the party as an excellent opportunity to introduce Romeo to some of the most beautiful young ladies of Verona, and perhaps then Romeo might realize that, in comparison to these 'swans', his Rosaline was but a 'crow'! Romeo protested that he couldn't possibly attend the party as he was a Montague. But Benvolio had thought of that: Romeo and his friend Mercutio, a kinsman of the prince of Verona, would go to the party disguised behind masks! Even though Romeo knew he would be in grave danger should he be discovered, he could not pass up the opportunity of being so close to his beloved Rosaline. So, the plan was agreed.

The party was a truly magnificent occasion, and, just as Benvolio had predicted, the ballroom was filled with the most beautiful young ladies of Verona. Soon one of them had caught the eye of Romeo. He was bowled over by her and could not restrain himself from blurting out to his friend how he had seen a girl of unparalleled beauty whom he intended to meet before the night was through.

Unfortunately, he was overheard by Tybalt, a nephew of Lord Capulet and cousin of Juliet, who recognized Romeo's voice and was outraged that a Montague should infiltrate the party of a Capulet under the cover of a mask. He would have killed Romeo there and then but was restrained by his uncle, who was aware of Romeo's good reputation around Verona and didn't wish to disturb the party.

Once the party was over, Romeo wasted no time in seeking out the young lady who had captured his heart and confessing his love for her. She too had been smitten by his looks, his gestures, and his words, and they talked animatedly together until she was beckoned away by her mother – who Romeo quickly discovered was none other than Lady Capulet! This was a disaster; how could he have fallen head-over-heels in love with a Capulet? But he was determined that not even this barrier would stand in his way. So that very night he sought out Friar Lawrence, a friend of the family (actually, a friend of both families), to marry him and Juliet the following day. After much chiding by the friar over Romeo's rapid change of affections from Rosaline to Juliet, he finally agreed to marry them, being persuaded not only by Romeo's obvious love for Juliet, but also by the hope that the marriage might heal the rift between the two families.

Early the next day, Romeo and Juliet were married, after which Juliet returned home to await the evening, when Romeo would come to the orchard as on the night before. However, later that same day, Benvolio and Mercutio ran into a group of Capulets with Tybalt at the head, and in the ensuing fight Mercutio was killed. At this point, Romeo, who was just passing by and so witnessed the death of his dear friend Mercutio, sought out Tybalt, and in the ensuing altercation Tybalt died.

News travelled fast, and soon the square was filled with the Lords Capulet and Montague and their followers and even the prince himself. Romeo's guilt was clear, as was the punishment of the law. But the prince had it within his power to pronounce a lesser penalty, should he be persuaded by all the facts of the case. He therefore pronounced that Romeo should not die but be banished from Verona for life.

When the news of the events reached Juliet, she was at first filled with rage against Romeo for slaying her precious friend Tybalt, but gradually her grief was turned to joy with the knowledge that her Romeo would at least live, though banished for life.

Meanwhile, Romeo, having escaped from the scene, sought refuge at the priory with Friar Lawrence. The friar was perplexed at the situation before him. When Romeo calmed down, it was agreed that he would take his leave of Juliet that night and travel to Mantua, where he would await the outcome of the friar's attempt to bring reconciliation to the families of the Montagues and Capulets, following which, the friar was in no doubt, Romeo would be pardoned by the prince.

These plans were, however, quickly wrecked by Lord Capulet announcing, within a matter of days after Romeo's leaving, that Juliet was to marry Count Paris, a kinsman of the prince! Juliet pleaded with her father that the time was not right to marry so soon after Tybalt's death – she dared not tell him she was already married! But her father would have none of it: Juliet was to marry the following Thursday.

Juliet, in great panic, sought the counsel once again of Friar Lawrence. After much consideration, the friar gave Juliet a drug which he instructed her to take on the evening before her wedding. The drug would put her into a deathlike sleep for forty-two hours, long enough to convince Count Paris that she was dead and to arrange for her removal to the family vault awaiting her burial. Here, Juliet would awake from her 'sleep' and escape to be with her beloved Romeo.

The bad news of Juliet's 'death' travelled quickly to Romeo – quicker than the messenger sent from Friar Lawrence with the real but secret news that Juliet's death and burial were a ruse and that they would be together soon. On receiving the bad news, Romeo was overcome with grief and not wishing to live without his Juliet, hurried back to Verona, purchasing, before he left, a vial of poison, which he would drink when beside her in the tomb. He arrived in Verona at midnight and was surprised to find Count Paris already at the tomb. Paris, assuming that Romeo, being a Montague, was up to no good at the tomb of a Capulet, accosted Romeo, and in the ensuing argument Paris was killed.

Juliet's tomb now contained alongside her the bodies of Tybalt and Paris; and it was here, by the side of his beloved Juliet – who was about to awake from her coma – that Romeo drank his own deadly poison. When Juliet awoke, saw the cup of poison in Romeo's lifeless hand, and heard the sound of people coming, she unsheathed her protective dagger and slew herself to lie forever beside her beloved Romeo.

When the prince and the Lords Montague and Capulet, heard the full story of Romeo and Juliet from Friar Lawrence, the prince rebuked the lords for their irrational enmities, and showed them what a scourge heaven had laid upon their offences. Accepting the prince's rebuke, their jealousies and hates were buried, and their long strife was laid aside.

—

Although the play has many subtleties of language and numerous subplots, the overarching theme is the destructive depth of hate between two important families. And although such hate could be swept aside on a personal level by the captivating love of the two young members of the two families; on a societal level, it could only be overcome by the two families recognizing that it was their feud which had brought about the tragic death of the two lovers.

10
The Merchant of Venice

Another story from the prolific hand of **Shakespeare**.

Jews, though expelled from many European cities in the Middle Ages, were a sizeable population in Venice, though still subjected to many forms of persecution and injustice.

Shylock was a Jew and a hard-nosed, well-known moneylender, who suffered much ridicule and persecution at the hands of Antonio, a wealthy Christian merchant of Venice, who hurled abuse at Shylock every time they passed on the Rialto.

One day, Antonio's beloved friend Bassanio, a noble Venetian but of little wealth, who regularly borrowed money from Antonio, asked Antonio for the rather large sum of three thousand ducats. He needed the money, he pleaded with Antonio, to allow him to woo Portia, whom he had known since childhood, and who had recently, upon the death of her father, become a rich heiress: she wouldn't possibly entertain him as a suitor without money. Antonio explained that he didn't have such a large sum available at the present time, but because of his great affection for Bassanio, he was prepared to ask Shylock, whom he despised, for the money. Antonio said it would be paid back from a large shipment of merchandise which even at that very moment was sailing towards Venice.

Shylock was rubbing his hands at the opportunity of exploiting Antonio's humiliation at having to come to him for a loan, even though Antonio told Shylock he would still despise him even if he offered him the loan: 'So offer me the loan as if I were an enemy and I will pay the extra terms.' But Shylock was determined to keep the upper hand, so offered the loan – but without any monetary interest; instead, he asked for a pound of Antonio's flesh! Antonio, out of his deep love for Benvolio and because of his confidence in the imminent arrival of his merchandise, accepted the terms, terms which he believed in any case Shylock had offered merely in sport, even though a bond had been signed.

Unfortunately, a short time later, Antonio received word that his ships had foundered, and all his merchandise was lost!

Shylock immediately took Antonio to court, refusing to accept full payment of the debt by Bassanio, who, using Antonio's loan, had successfully wooed the beautiful heiress Portia, and thereby gained access to her great wealth.

At the court, all seemed lost for Antonio until the arrival of an unknown, very skilful lawyer – who was actually Portia in disguise! She pointed out to the court that it would be impossible for Shylock to take his pound of flesh, which Shylock repeated again and again was absolutely within the letter of the law, without shedding Antonio's blood, the blood of a Christian citizen, which, by the letter of the same law, would mean Shylock would forfeit his lands and all his goods to the state of Venice!

With this revelation, Shylock quickly changed his mind, and said that instead of the pound of flesh he would accept the money previously offered by Bassanio.

But Portia again intervened, saying that Shylock had adamantly refused the money earlier, insisting, vengefully, on his pound of flesh;

that that was what the bond had specifically stated; and that that and only that was what should be granted. But be aware, she continued, that by that same law, not so much as a single drop of blood could be shed, nor an excess of flesh of even the weight of a hair could be carved, otherwise not only would his lands and goods be taken, but also his life. At this, Shylock, in great panic, quickly said, once again, that he would take the money offered by Bassanio.

But once again, Portia interrupted, pointing out to the court that Shylock had conspired against the life of one of its citizens, and only the duke of Venice, presiding over the court, could save him from the due penalty of the law, by exercising Christian mercy. Upon hearing this, Shylock fell to his knees before the duke pleading for his mercy. The duke duly offered Shylock his life, but ruled that his wealth be confiscated, with half awarded to Antonio and the other half to the State.

Antonio, showing great generosity, then proposed to Shylock that he would forfeit his share, on condition that Shylock sign a deed that he would, upon his death, hand over the same amount to his daughter and her husband, whom Shylock had disinherited upon their marriage, because her husband was a Christian.

—

Again, as with all Shakespeare's plays, there are many sub-stories in *The Merchant of Venice*, but its dominant theme is the destructive consequences of excessive greed and hatred of another's creed. The constructive route, making possible the continuance of life and with it the possibility of a change of heart, lies in offering generosity and mercy.

11
Mahatma Gandhi

Mahatma Gandhi was born in a small village in India, under the British Raj, in 1869, and died at the hands of an assassin in New Delhi in 1948. He is popularly known for bringing about the independence of India from British rule in 1947 through his absolute commitment to non-violent activism. His example has inspired civil rights movements around the world. His non-violent protests on many issues of oppression of minorities led to him being arrested and jailed on numerous occasions. He also embarked, on three critical socio-political occasions, on hunger strikes. He suffered regular beatings at the hands of the authorities and was subjected to cutting and vile personal criticism. The criticism was particularly vicious from the British political leaders of the day.

Gandhi began to form his philosophy of life while practicing law in South Africa in his early twenties, having studied law at University College, London, and being 'called to the bar' in 1891. His philosophy was founded on his deep inner awareness of a unifying 'life-soul' common to all humanity. He found this belief to be articulated primarily in his beloved Hindu scriptures but also in the holy texts of all the world's major religions. For him, 'God has no religion … [and Jesus] belongs not only to Christianity but to the entire world, to all races and all people.'

From this unshakeable inner awareness sprang his **satyagraha,** his modus operandi, his activism, his total dedication to the pursuit of truth. He wrote: 'God is truth. The way to truth lies through *ahimsa* [non-violence], *satyagraha*.' And since he believed we all possess a 'soul force', a 'God-in-us' presence, we must a priori treat all people equally – and this applies to those who would see themselves as our enemies as well as our friends: 'It is easy enough to be friendly to one's friends. But to befriend the one who regards himself as your enemy is the quintessence of true religion. The other is mere business.' And the path to follow in the resolution of differences is the path of non-violent activism: 'Non-violence is not a garment to be put off and on at will. Its seat is in the heart, and it must be an inseparable part of our very being.'

He lived out his life at the level of survival sufficiency and protested against inhumanity, marginalization, and oppression, wherever it occurred. He criticized Western civilization as one driven by 'brute force and immorality', contrasting it with his view of Indian civilization as one driven by 'soul force and morality'.

—

From his very humble beginnings, through a life of humility and fearless non-violent confrontation of oppression, injustice, and exploitation of every kind, he became a deity, mahatma ('great soul'), among his own people and revered throughout the world.

12

Martin Luther King

'I have a dream,' said **Dr. King, the American civil rights leader,** in his most famous speech, 'that my four little children will one day live in a nation where they will not be judged by the colour of their skin but by the content of their character. I have a dream today … I have a dream that one day this nation will rise up and live out the true meaning of its creed: "We hold these truths to be self-evident, that all men are created equal." I have a dream …'

Martin Luther King Jr. was a black Baptist minister and social activist who led the civil rights movement in America during the 1950s and 1960s until his life was tragically cut short at the age of thirty-nine by an assassin's bullet.

He promoted non-violent resistance against entrenched prejudices towards black people in the Southern states, finally bringing about the end of the policy of segregation and the systemic oppression of non-white people in the Southern states.

His non-violent activism first came to prominence when he was chosen to head the Montgomery Improvement Association, formed by the black community to bring about desegregation of the city buses in Montgomery, Alabama. It followed the arrest of Rosa Parks, a black woman, who refused to vacate her bus seat in favour of a white man. As a result of a sustained campaign of non-violent action, led by King, the U.S. Supreme Court ruled in late 1956 that segregation on buses was unconstitutional.

But this was only the beginning for King. He went on to lead campaigns against segregation and other racial practices in other sections of society across the United States. He was jailed on several occasions, but he recognized from the beginning that that was the price he would have to pay. His most notable campaigns were Birmingham, Alabama, in 1963, culminating in a historic march to Washington in August 1963 where he delivered, from the steps of the Lincoln Memorial, to a crowd of some 200,000, his 'I Have a Dream' speech, and which led to the Civil Rights Act of 1964; Selma, Alabama, in 1965, leading to the Voting Rights Act of 1965; and in Memphis, Tennessee, in 1968, in support of a strike by sanitation workers.

The day after delivering his 'I've Been to the Mountaintop' speech, he was shot dead by James Earl Ray on the balcony of his hotel.

—

Martin Luther King was a man of great courage, who suffered many indignities, was subjected to much personal violence, was jailed on several occasions, and finally accepted the definite possibility of assassination, because he believed in equality, justice, and freedom from oppression for all people, regardless of the colour of their skin, their ethnicity, their creed, or their position in society.

13
Nelson Mandela

Nelson Mandela was another political leader and social activist but this time in South Africa.

He spent twenty-seven years of his life in prison for opposing the governing political system of apartheid introduced by the white government of South Africa in 1948 – though this specific policy was the culmination of generations of ad hoc interventions through laws and practices at both local and national level. Apartheid literally means 'apart-hood' or 'apartness' and was the systematic ordering of all aspects of society – political, economic, and social – according to race, consolidating, by law, total white supremacy.

Mandela was born in 1918 in a small village in the Eastern Cape. His father was principal counsellor to the acting king of the Thembu people. He began studying for a Bachelor of Arts degree at the University College of Fort Hare but was expelled for joining in a student protest. He belatedly completed this degree after spells of employment, in 1943. He also began studying for a law degree when doing his articles for a firm of attorneys. Once again, he did not complete his degree, but on the basis of a two-year diploma in law coupled with his BA, he was allowed to practice law, and in 1952, along with Oliver Tambo, established South Africa's first black law firm.

Mandela's activist work ran alongside his legal work and his studies basically from 1944 when he joined the African National Congress (ANC). Apart from what he perceived as the daily general oppression of the black peoples of South Africa, his main thrust was to bring an end to the apartheid system in favour of a non-racial Constitution. He stood trial and was jailed on a number of occasions, once for treason, and finally for sabotage in what became known as the Rivonia Trial in 1964. He defended himself at this trial and his closing speech from the dock, facing the death penalty, has become famous throughout the world: -

> I have fought against white domination, and I have fought against black domination. I have cherished the ideal of a democratic and free society in which all persons live together in harmony and with equal opportunities. It is an ideal which I hope to live for and to achieve. But if needs be, it is an ideal for which I am prepared to die.

On 11 June 1964 Mandela and seven others accused were convicted and sentenced to life imprisonment and sent to Robben Island.

In prison he continued to make the case against apartheid and lived out his belief that all people should be treated with dignity and respect – even the prison warders. From the prison he won the moral argument to such an extent that even before his release in 1990 he was being consulted by the presidents of South Africa, first P. W. Botha and then F. W.de Klerk.

In 1991 Mandela was elected ANC president, to succeed his ailing lifelong friend, Oliver Tambo. And in 1994 he was inaugurated as South Africa's first democratically elected president.

In 1993, in recognition of the courageous steps taken by each man, F. W. de Klerk and Nelson Mandela were jointly awarded the Nobel Peace Prize.

—

Nelson Mandela, despite all the difficulties placed in his path, never wavered in his belief that all human beings, regardless of the colour of their skin or their religion or their culture or their place in life, should be treated with dignity and respect and afforded equal opportunities and equal treatment before the law. Though oppressed by the colour of his own skin, he never answered racism with racism. Hence the South Africa he tried to build has been called the Rainbow Nation.

14
The Good Samaritan

Jesus of Nazareth told many parables. He was very adept at explaining what he wanted to say using stories drawn from the everyday experiences of his listeners. The story of the good Samaritan is probably his most famous.

He told this story to answer a question put to him by a persistent, skilful lawyer (the law under which the people lived at the time was the Holy Law of Judaism). The lawyer, probably wanting to show how little Jesus knew of the law, asked Jesus what he had to do, the way he should live, what his priorities should be, in order to inherit eternal life.

The answer Jesus gave was in two parts: the first was to ask the lawyer what Moses's law, which the lawyers all purported to live by in every detail, said about the matter. Without hesitation the lawyer replied: 'You must love the Lord your God with all your heart, and with all your soul, and with all your strength, and with all your mind. And you must love your neighbour as you love yourself.' 'Well said,' Jesus replied. 'Live like that and you will have eternal life.' But the lawyer was determined to catch Jesus out, so he asked him: 'And who is my neighbour?'

This led to Jesus offering the second part of his answer; he told the lawyer a parable. There was a man, a Jew, making the dangerous journey from Jerusalem to Jericho, about twenty-four miles away. The pathway was down a very steep, rock-strewn gorge, making it an ideal location for bandits and robbers. And this was exactly what happened to the Jew; he was attacked, robbed, and left for dead by the side of the track.

Presently, a priest came along, who, like the lawyer who asked the question, lived his life strictly according to the Holy Law. When he saw the man lying by the side of the track, he stepped around him onto the opposite side of the track.

Shortly afterward, a Levite, who had a special role in all matters relating to the holy temple, the house of God, in Jerusalem, came along. When he saw the man lying there, he went over to take a closer look, but then he too continued on his way.

Finally, a Samaritan came along. Now Samaritans and Jews despised each other: the Jews believed *they* were the true descendants of Abraham, and the Samaritans were not, because they were of mixed blood due to intermarrying with other peoples during the period of the Exile. They were tainted; they were not pure descendants. Therefore, the Jews and the Samaritans had nothing to do with each other. Yet it was this man, the Samaritan, who went over to the stricken Jew, tended to his wounds, lifted him onto his donkey, and took him to the nearest inn. He even gave the innkeeper money in advance and told him to do whatever was necessary for the injured man, and he would pay him any extra he required on his return journey.

When Jesus had finished the story, he turned to the lawyer and asked him which of the three people he would class as the 'neighbour' to the man who had been attacked? The lawyer replied that it would be the one who had shown mercy. 'Then you go and do the same,' said Jesus.

—

The question we all need to find the answer to if we are to lead full, moral lives which enrich not only our own lives but also the lives of others, is: Are there any who live outside our mercy? Of course, we must live justly within the law, but that justice must always be tempered with compassion and mercy, no matter who is involved.

15
The Lost Son

This is another well-known story from **Jesus of Nazareth**.

A father had two sons, both of whom would inherit his land, livestock, and money. One day, the younger son went to his father and asked if he could have his share of the inheritance immediately. His father, though probably very taken aback by this request, agreed.

Within a few days the younger son gathered all his possessions together and with his inheritance tucked into his belt, left to travel the world. He had a good time and went through his inheritance at quite a rate. Unfortunately, just at the time when his money was running out, the country he was in was hit by a very severe famine. He was struggling to survive, but eventually managed to find a job on a pig farm, feeding the pigs. But food was still very scarce, and before long he began to envy even the food the pigs were eating – and still no end to the famine was in sight.

Starving as he was, he got to thinking about back home on his father's farm. There everyone, including the hired hands, had more than enough to eat. So, it occurred to him that he could ask his father to take *him* on as a hired hand, as he knew he could not expect to be treated as a son again. With this thought in mind, he set off back to his father's farm.

As he was approaching the farm, his father caught sight of him in the distance and ran toward him with his arms outstretched. The son hurriedly blurted out his prepared speech: 'Father, I know I've done wrong, made a mess of things and don't deserve to be called your son, ….' His father would have none of it; as far as he was concerned, this was his son whom he had thought he had lost forever. They would have a big celebration in his honour – even killing the fatted calf for the feast!

The elder son, however, was not so pleased, and refused to join in. His father pleaded with him, but he would not listen, saying that he had faithfully served on the farm over all the years, yet he had not been given even so much as a kid goat to celebrate with his friends, let alone the best fatted calf!

His father said to him: 'My son, you are always at my side. Everything that's mine is yours. But we just had to celebrate, because I thought this brother of yours was dead, but he's come back to life; he was lost and now he's found.'

—

This is a powerful story about how repentance can lead to a renewal of life; and how forgiveness has the power to heal relationships.

16

The High-Ranking Religious Professional and the
Looked-Down-Upon Tax Collector, at Prayer

Jesus drew on his observations of everyday life for his parables. Here is a hard-hitting parable of two men at prayer in the temple.

Two men went up to the temple to pray, one a respected Pharisee, a scholar of the Law: the other a tax collector for the occupying power, the Romans.

At prayer, the first man stood in a central place for all to see and prayed: 'I thank you, God, that I am not like everybody else, thieving, unjust, adulterous, and especially not like that tax collector over there. I fast twice a week, and I give tithes on everything I earn.'

Meanwhile, the tax collector, who was disliked and shunned by the public because he worked for the despised Romans, stood off to the side out of public view, beat his chest, and prayed: 'God, have mercy on me, sinner that I am.'

Jesus turned to his hearers and commended the second man to them, saying that it was he, the tax gatherer, who would go home with the blessing and not the first.

—

This graphic little story speaks of the dangers of self-righteous pride and commends the virtue of humility.

17

The Unforgiving Servant

Not perhaps the most popular of Jesus's parables, but a telling one, nonetheless.

A king decided one day it was time to settle his accounts with his servants. In the process a servant came in who owed the king a very large sum of money. There was no way the servant could pay off the debt, so the king ordered that the servant, along with his wife, children, and goods, should be auctioned off at the slave market.

At this prospect, the servant threw himself at the king's feet and begged again and again for time to pay. The king was moved by his plea, and being a man of compassion, the king cancelled his servant's debts and sent him back to work.

Soon after returning to his duties, the servant came across a fellow servant who owed him a small amount of money. He grabbed his fellow servant by the scruff of the neck and demanded he repay the debt immediately. The servant threw himself down before him and begged for time. But the lender wouldn't listen to him: he had the borrower arrested and put into jail until he had paid off his debt in full.

When the other servants heard what had happened, they immediately reported it to the king. On hearing it, the King demanded that the first servant be brought back before him, and said to him: 'You evil, callous man, you owed me a huge sum of money, but out of compassion I let you off, yet now you have treated this man harshly for only a trifling sum. Why couldn't you have shown the same compassion for him as I did for you?' And with that the king had his servant put in jail until he had paid back the entire sum.

—

This story underlines the importance of compassion and fairness in human relationships, summed up perhaps by what is commonly known as the Golden Rule: 'Do unto others as you would have them do unto you.'

18
The Resourceful Servant

A wealthy man was going away on a long trip, so he called his servants together, distributed his wealth among them, and told them to look after things until he got back. To the first servant he entrusted thirty thousand silver coins, to the second twelve thousand, and to the third six thousand, according to how effectively he thought each of them would manage the money.

After a long absence, the master returned and asked for an account from each of his servants.

The first servant reported that he had made another thirty thousand silver coins. The master congratulated him and invited him to join his celebrations.

The second servant reported that he too had doubled what the master had left with him. The master congratulated him too and invited him to join his celebrations.

The third servant, however, could only report a return of the same amount he had been given, because he had not done anything with it but simply kept it safe. The master was not pleased, saying to the servant that the least he could have done was invest the money safely with low interest. Since he had produced nothing, he would not be invited to the celebrations and the money would be taken from him and given to the first servant.

—

We are not all born equal. We have different abilities and opportunities, but whatever we have we should try to use to the full.

19
The Farmer Sowing His Seed

A farmer went out to sow his seed.

Some of it fell on the pathway, where it was trampled under-foot, or the birds got it.

Some of it fell on stony ground where there wasn't much soil, so it could hardly take root and quickly withered away in the hot sun.

And some of it fell among the weeds which smothered it, so it produced no crop.

But some of it fell on rich earth which produced a bumper crop.

—

We must not be discouraged from presenting the good life, even though sometimes it may appear to be a waste of time. We must always keep sowing, always strive to promote human flourishing.

20
Pithy Sayings of Jesus

Jesus was well known in his day for what we would call today his 'one-liners': that is, very pithy sayings containing some usually hard-hitting truths. His listeners were usually on their own to work out the meaning and to whom they were directed. And the same applies to us today; we can ignore them, or believe they apply to somebody else. Or we can absorb them and allow them to refine the moral compass of our souls.

It is written, 'Human beings are not to live on bread alone,
But on every word that comes out of God's mouth.'

There is more to living than food and clothing.

The Sabbath was made for man,
Not man for the Sabbath.

Enter by the narrow gate:
For the gate is wide and the way is easy,
That leads to destruction,
And those who enter by it are many.
For the gate is narrow,
And the way is hard,
That leads to life,
And those who find it are few.

Some who are first will be last,
And some who are last will be first.

Love your enemies and pray for those who persecute you.

Forgive and you will be forgiven.

You have heard that it was said 'An eye for an eye, and a tooth for a tooth',

But I say to you, don't react violently against the one who
is evil:
When someone slaps you on the right cheek, turn the
other as well.

When you are about to appear with your opponent before
the magistrate,
Do your best to settle with him along the way.

You see the sliver in your friend's eye,
But don't see the timber in your own eye.
When you take the timber out of your own eye,
Then you will see well enough to remove the sliver from
your friend's eye.

Be on your guard against the scholars
who like to parade around in long robes,
and who love to be addressed properly in the marketplace,
and who prefer important seats in the synagogues
and the best couches at banquets.

When you give to charity, don't let your left hand
Know what your right hand is doing.

Whoever has two shirts should share with someone who
has none.
Whoever has food should do the same.

If you have money,
Don't lend it at interest.
Rather give it to someone
from whom you won't get it back.

It's easier for a camel to squeeze through the eye of a
needle,
Than for a wealthy person to get into God's domain.

No servant can be a slave to two masters.
He will either hate one and love the other,
Or be devoted to one and disdain the other.
You can't be enslaved to both God and a bank account.

Whoever tries to hang onto life will forfeit it,
But whoever forfeits life will preserve it.

Don't hide your light under a basket,
But put it on a lampstand, where it sheds light for all.

You'll know who folks are, by what they produce.
Since when do people pick grapes from thorns or figs from
thistles?

Let the children come up to me, don't try to stop them,
After all, God's domain belongs to people like that.

21
'Do It Anyway'

These words were pinned on the wall of Mother Teresa's Children's Home
in Calcutta. They are an adaptation of 'The Paradoxical Commandments'
(or the 'Do it Anyway' version of Dr. Kent M. Keith's 'Paradoxical
Commandments').

People are often unreasonable, illogical, and self-centred.
   Forgive them anyway.
If you are kind, people may accuse you of selfish, ulterior
motives.
   Be kind anyway.
If you are successful, you will win some unfaithful friends
and some genuine enemies.
   Succeed anyway.
If you are honest and sincere, people may deceive you.
   Be honest and sincere anyway.

What you spend years creating, others could destroy overnight.

Create anyway.

If you find serenity and happiness, some may be jealous.

Be happy anyway.

The good you do today, will often be forgotten.

Do good anyway.

Give the best you have, and it will never be enough.

Give your best anyway.

In the final analysis, it is between you and God.

It was never between you and them anyway.

21

Well-Known Proverbs

Proverbs are the distillation into a few words of many centuries of proven human experiences. There is nothing scientific about them; they are simply the common-sense understandings of the general outcomes of actions and happenings experienced and practiced by human beings over many centuries. They survive because each of them contains a useful kernel of truth: they represent what happens, not in every case, but by and large. The fact that we still use them today is testimony to their positive contribution to human flourishing.

Proverbs then, are very useful to the nurturing, growth, and development of the moral compass of the soul:

Actions speak louder than words

A journey of a thousand miles begins with a single step

All that glitters is not gold

A stitch in time saves nine

Beauty is in the eye of the beholder

Better late than never

Don't count your chickens before they hatch

Don't judge a book by its cover

Don't put all your eggs in one basket

Don't put off until tomorrow what you can do today

Fortune favours the brave

Good things come to those who wait

Hear no evil, speak no evil

Honesty is the best policy

If at first you don't succeed, try, try, and try again

If you play with fire, you will get burnt

Knowledge is power

No man is an island

Nothing ventured, nothing gained

People who live in glass houses should not throw stones

Practice makes perfect

The early bird catches the worm

The grass is always greener on the other side

There is no time like the present

Two heads are better than one

Two wrongs don't make a right

When in Rome do as the Romans do

When the going gets tough, the tough get going

Where there's smoke, there's fire

Where there's a will, there's a way

You can lead a horse to water, but you can't make it drink

—

All these understandings will achieve nothing if they are kept in cotton wool. If our lives are closed down, if they are moribund, if we have become cabbages, if we have become complacent, if we have become resigned to our situation, we need a way to restart the process which leads to flourishing. And all it demands is the courage to take one step, one selfless act, however small – or large.

What follows is the story from the Book of Genesis of a man who did just that.

22
The Story of Abraham

For most of us uprooting our lives, leaving our families, and moving home because of the necessity of earning a living, is bad enough; but I wonder how many of us would have the vision and courage to do it simply because we felt an inner urge to go? This is what Abraham did.

For him, it was the inner 'voice' of his God that told him to uproot everything, everything a very successful man had – his beautiful wife, his loyal nephew, his gold and silver, his herds of cattle, his slaves – and head off on a journey to a place which would only be revealed to him along the way! He didn't know when he set out just where he would end up, but he had the promise from his God: 'If you do this, I will make you the father of a great nation.'

So, Abraham left his friends and very successful life behind and headed north, following the river Euphrates. His journey took many twists and turns before he reached his final destination at Hebron in Canaan. One such twist was having to divert down into Egypt because of a famine in Canaan. On entering Egypt, his wife had to pretend to be his sister, so that Abraham, a foreigner, wouldn't be killed and his possessions confiscated, as was the practice of the day. Unfortunately, this ruse led to his wife being taken, as his sister, into the king's harem! All Abraham could do was stand by and watch and keep his head down. Eventually, the king discovered the truth, and Abraham and his wife were banished from Egypt, so they headed north, back to Hebron.

Abraham endured all these things because of an irresistible inner compulsion – for Abraham, the persistent inner 'voice of God' – to leave his comfortable existence and step out into the unknown. But the outcome was that he did indeed become the 'father of a great nation' as his God had promised: that nation was the Hebrews, the people of Israel, or the Jews as we know them today.

—

The story of Abraham reminds me of one of those proverbs: 'Nothing ventured, nothing gained'. Yes, it takes courage to step out into the unknown, but from the very first moment of standing on our own two feet as tiny children or riding our bicycle unaided for the very first time, stepping out like this has led to a huge expansion of our human experience and, through that, to a great enrichment of our lives.

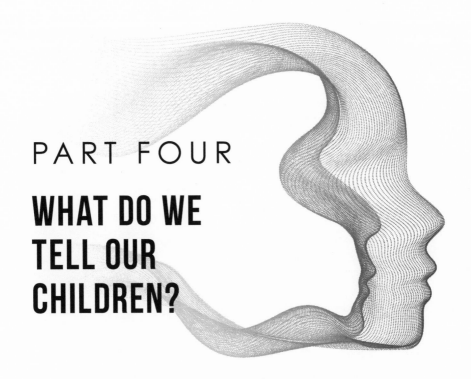

PART FOUR

WHAT DO WE TELL OUR CHILDREN?

There is no need for temples,
No need for complicated philosophy.
Our own brain, our own heart is our temple:
The philosophy is kindness.

The Dalai Lama

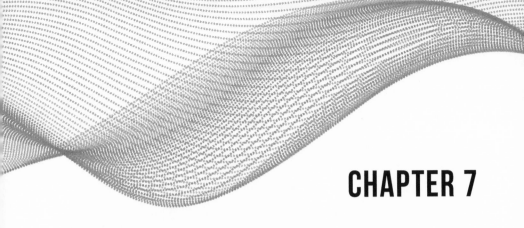

CHAPTER 7

Moral Stories for Children

Perhaps you're not into the theology or philosophy of the earlier parts of this book. That's all right, it's not everyone's cup of tea. It's certainly not, in the case of children! They simply like a good story for its own sake – particularly if it involves animals!

I've tried my hand at telling such stories over the years. My favourites recount the adventures of two animals, an elephant called Jimbo and a bird called Conché. And four or five years ago I published them under the title 'Climbing the Stairway to Heaven'. What follows in this chapter, are some of the stories from that publication.

The advantage of children's stories is that personification of the animals in the stories is received by the children as if it were normal. They live quite happily in a world where everything, animate or even inanimate, is just like them. So, the fact that the animals in the stories feel our emotions and speak our language is not a problem for them.

Using animals also has a huge advantage for adults. It enables us to contemplate concepts and virtues without any of the historical formulations surrounding them bequeathed to us by our ancestors. In other words, animal stories allow the demystification of the accumulated religious concepts of the human species and allow us to interpret the world of present experience with our current understandings – not historical or inherited ones.

So, embedded in these stories of Jimbo the elephant and Conché the bird are many of the main themes of human relationships – love, forgiveness, hope, patience, self-control, thankfulness – and the wider, deeper themes of life – its meaning and purpose, and its powerful emotions of sadness and happiness.

1

A Nose for Someone in Trouble

It was a bright sunny morning with cloudless blue sky framing the leaf-canopied treetops. Conché looked out from her night-nest and trembled with excitement – another new day! *What shall I do today?* she wondered, as her friends began to wake around her and shout their joyful news at the dawning of another new day. Soon, the air was filled with a cacophony of sound as the whole forest stirred to life once more.

Conché stretched her toes and gently flapped her wings, allowing the warmth of the sun to seep into her bones. Before long, her friend Chippy arrived: 'We're thinking of going to Big Roller today, Conché. Want to come?' Conché's heart leapt: Big Roller! Would this be the day? she wondered. Would she finally manage to beat all the others? Would she be the first today, the fastest, the best?

'Count me in,' called Conché excitedly, as she sprang off her bed to join her friends circling high above in the clear blue sky.

Off they flew over the treetops of the great expansive forest, riding the gathering thermals wherever they could to preserve their energies for the race that was to come. Suddenly, the whole flight swooped down as one, to perch in the top branches of the biggest, tallest tree in the forest.

What a tree! It was huge! It stood like an armoured giant keeping guard over the whole forest. This was Big Roller! But it wasn't just its size which gave it its name; it was also the way it held itself. Its branches thickened and stretched wider and wider from top to bottom; and the spaces between the branches took up such unique dimensions as to create a hair-raising, jaw-dropping roller-coaster ride for any bird who had the courage to fly it!

Would Conché have the courage today? Would she be able to fly fast enough, weave accurately enough, turn sharply enough? The consequences of only the slightest misjudgement would be disastrous. Not for nothing had their parents branded this tree the 'Tree of Death'!

But that warning was far from Conché's mind today. Today would be her day – she could feel it in her bones! Yes, she was terrified; yes, her limbs were shaking; but when it was her turn to fly Big Roller, she would put everything she had into it. She would twist and turn through the branches like a moth dancing around a flame. And she would be quick – oh, she would be very quick. No one would beat her today!

At last, it was her turn. The others had been good – some very good – but Conché was sure their times were beatable.

She launched herself away and quickly stretched her wings to the limit. She dived with all her courage round the first branch, then immediately adjusted herself to skim under the second, just brushing her head-plume on the underside bark. After a quick check on her upward roll before swooping again, this time to scrape the underside of her breast as she rounded a thicker branch. 'Still three more to go,' she whispered to herself as she twitched her tail to correct her line for the next branch. 'Too fast, too fast!' she shouted, as the huge shape sped towards her. Her heart was in her mouth as she frantically dropped her wingtips in a desperate effort to slow herself down. 'Still too fast, too fast!' she yelled, knowing full well that at this speed even if she miraculously navigated around this branch, she had no hope for the next one.

And that's how it happened. It was an agonizing thump: the air was driven out of her chest, her whole world blacked out, and she dropped like a stone to the forest floor.

When eventually the light began to return, she felt awful – her head was throbbing, her eyes were unfocused, and her chest felt compressed as if under a heavy weight. Slowly she checked her limbs; left leg – okay; right leg – okay; left wing – okay; right wing – she couldn't move her right wing! She tried again, but the pain was excruciating. Now fear welled up within her as she realized that her wing was broken, and she couldn't fly!

'No! No! No!' she cried, as panic began to sweep over her, 'What am I going to do? I'm going to die!' Conché was only too aware of how dangerous it was for a little bird on the forest floor. She knew that down there she was no match for most of the animals who roamed the forest; even those who weren't interested in eating her for dinner would have no regard for her as they padded their way along the crisscrossing pathways of the forest.

'Why was I born so small?' she sobbed. 'Why wasn't I born a tiger or a lion or an elephant?'

The shifting shadows as the sun moved on its unstoppable course, and the slight rustling of leaves caught by a warm mould-filled breeze, all added to the sense of fear and foreboding which, in wave after wave, overwhelmed Conché and reduced her to a tearful, quivering little mass on the forest floor.

Suddenly, she felt the ground begin to shake beneath her, only a little at first but slowly increasing until she was almost bouncing! Out of the corner of her eye she caught a glimpse of the underside of a huge foot. It was an elephant – and it was about to step on her!

Conché let out a piercing cry and screwed her eyes tightly shut. Her heart was in her mouth, and she could hardly breathe as she waited for the huge foot to come thumping down and squash her … But nothing happened! No foot! No squashing! Nothing!

Instead, a big booming voice called to her: 'Now then, little bird, what are you doing down there? That's a very dangerous place for a little bird like you.'

The elephant was young Jimbo, out on his afternoon stroll. He'd spent the morning playing with his brothers and sisters of the herd and now he just wanted a quiet little walk in his favourite part of the forest. He loved the forest, with all its smells and sounds and wonderful trees. But now, today, he had come across something unusual – a little bird that wouldn't or couldn't get out of his way. *What should I do?* he thought to himself. *Oh, it's only a little bird and there are thousands of them in the forest. If it doesn't get out of my way, so be it.*

But this little bird was crying from the bottom of its heart … so he thought he'd stop and find out why. Through deep sobs the little bird fearfully croaked her story: 'You see, Mr. Elephant, I was playing with my friends in the trees; it was a game to see who could go the fastest and I … I went too fast … and couldn't get round the bottom branch. and I hit it very hard and – and I knocked myself out and ended down here … and I've broken my wing and can't fly. And I don't know what to do … and I'm very frightened.'

'Ummmm,' murmured Jimbo, as he took a careful step backwards.

Before Conché realized what was happening, a big, long, grey nose landed with a thump on the earth beside her. And the booming voice of Jimbo rang out again: 'Hop onto the end of my nose, little bird.' Conché was in a panic and didn't know what to do. What was the elephant's intention? Could she trust him, or was he simply going to toss her away? With her heart pounding, Conché plucked up all her courage – and some! – and hopped onto the end of Jimbo's nose.

Very, very, carefully, Jimbo swung his nose high into the air and dropped it gently onto the top of his very large head, between his huge floppy ears. 'Now then, little bird', said Jimbo, 'hop onto the top of my head; you'll be quite safe there, and you can stay until your wing is healed.' And with that, Jimbo set off again on his afternoon walk.

For many, many years Jimbo and Conché remained very good friends and enjoyed many adventures together … and it all began with a nose for someone in trouble!

2

Stuck in a Hole

There was nothing Jimbo liked better than to stroll through the forest on a lovely sunny afternoon. Maybe it was the smells he found so evocative – the cool, musty smell of damp earth and rotting vegetation; the pure aroma of spring blossoms; the tangy sap leaching from broken branches. Or perhaps it was the changing shades of light as the sun pursued its heavenly course. Or the light breezes which swayed the branches and fluttered the leaves. Maybe.

But for Jimbo there was now something more precious than all of these: it was the glorious singing of the birds. In fact, one bird in particular, Conché, his very best friend. From her he would hear the most beautiful songs, sung over and over again, just for him!

Every afternoon, she would swoop down and perch on Jimbo's head, between his huge floppy ears, and sing, and sing, and sing. They had been friends since the time Jimbo had rescued her from the terrifying dangers of the forest floor, after she had broken her wing trying to break the record for flying the Big Roller.

For Conché too, this was her favourite time and her favourite place, bobbing merrily on the top of Jimbo's head, singing to her heart's content. She knew Jimbo loved to hear her sing, so she sang and sang.

But today was not going to be like every other day. Suddenly, Conché felt herself dropping! Instantly, she flapped her wings wildly and, with heart pounding, rose swiftly into the sky. She looked down, but she couldn't see Jimbo anywhere! Where was he? What had happened? 'Oh my! Oh my!' she cried, 'He can't just have disappeared – elephants can't do that!' But where was he? 'This can't be happening', she cried out in despair.

Then she heard a loud trumpeting call, and she knew it was Jimbo, and she knew he wasn't very happy! But where was he?

Then she spotted him. He was at the bottom of a deep hole in the forest floor! The hole had opened up under Jimbo's weight, and now he was stuck. No matter how hard he tried, he couldn't get out; the sides were too steep, and there was nothing he could grab with his trunk to haul himself out.

He struggled and struggled, trying to climb the sides of the hole, but it was no use. He was stuck! And he was exhausted! Even his angry bellows had grown weaker and weaker, until he could hardly be heard. He stood, in the bottom of the hole with his head bowed low, his trunk sagging between his legs and his ears drooping like wilted flowers, feeling very, very sorry for himself!

As the day wore on, with his hope almost gone, he began to feel irritated by flurries of leaves and twigs which kept falling onto him. He constantly shook them off with a shake of the head and a shrug of the shoulders – then he stamped his feet in annoyance!

Conché had at first panicked when she saw Jimbo's terrible situation. She could see that there was no way he could get himself out of the hole. But what could she do? How could *she* help Jimbo get out of the hole? She was only a tiny bird, and he was a great big elephant! But she just couldn't just leave him in the hole to die! Her brain was in turmoil as she flew round and round the hole, hearing the plaintive cries of her friend, but not daring to fly down into the deep darkness of the hole.

Suddenly, an idea flashed into her mind, and the more she thought about it, the more she came to believe that it would work.

She flew off, rising high above the trees, trilling a call which she knew her friends would recognize as an emergency call. And sure enough, they did and quickly joined her in her flight. Before long there were ten, then twenty, then forty, then hundreds of her friends, forming a vast flock, like a swarm of bees following a new queen, flying with Conché until she brought them to rest in the branches of their favourite tree, Big Roller.

There they listened in grave silence to Conché's sad and upsetting story. She finished by pleading for their help to free her friend Jimbo. The silence was intense, no one could think of anything to say. Finally, a deep voice from somewhere in the lower branches, broke the silence: 'Conché', the voice said, 'you know that we would all love to help, because we love Jimbo too, but there is no way we can help Jimbo out of that hole: he is so big, and we are so small.'

'No, no', cried Conché. 'There is a way if you'll all help me.' And she began to outline her plan.

And that is why it was that Jimbo, stuck in his deep hole, kept feeling those irritating showers of leaves and twigs falling on him! All afternoon it went on! Shower after shower of leaves and twigs! He was so annoyed, he would shake his huge head and shrug his broad shoulders, and stamp his big, padded feet. Over and over again, he shook his head, shrugged his shoulders, and stamped his feet!

Slowly, very slowly, Jimbo began to realize what was happening – he was rising! His hole was being filled with twigs and leaves! And he was standing on the top of them! All the birds of the forest were picking up the leaves and twigs and whatever else they could carry and dropping them into the hole where Jimbo was stuck!

Before the sun went down, the hole was full, and Jimbo was able to clamber out and rejoin his overjoyed friends!

Over the years that followed, whenever Jimbo remembered that day, he was overcome with emotion at the love of his friends, who had pulled together, against almost impossible odds, to rescue him from a very frightening, life-threatening, situation.

3
The Log Den

'What about building a den?' suggested Jimbo one day, as he and his friends stood messing around, waiting for the rain to stop.

'What do you mean?' queried Jondle. 'Somewhere to shelter on our own, where we can do what we like?'

'Something like that,' replied Jimbo.

'I'm in,' said Drun.

'Me too,' said Trindle eagerly, casting an adoring look at Drun.

'No, no,' said Jondle. 'If we build a den, there should be no girls allowed!'

'Then you can count me out,' said Jimbo quietly. 'Either we're all in or none at all.'

There was an awkward silence. Some were kicking their feet in the earth, and Drun was casting sly glances at Trindle, who was looking very glum with her head bowed.

'Oh, all right then,' said Jondle. 'Anything's better than standing about here doing nothing.'

'Okay, then,' said Jimbo. 'First, we need to find a clump of trees so arranged that we can slot logs for the side walls between them – not too close together and not too far apart.'

They trooped off into the forest to search for the best arranged clump.

They searched and searched, until eventually Trindle shouted, 'Over here!' They all swung over to where Trindle was standing.

Jimbo looked around and then paced the distance between the trees. 'Yes', he said, 'I think these'll do. What do you think. Jondle?'

'Mm, I suppose so,' said Jondle grudgingly.

'Okay then', said Jimbo, 'we need to gather some decent logs to form the back wall: they need to be able to slot between these trees here.' He pointed to where the back wall would be. 'Then we will have to bring them out this way' – he gestured with his trunk – 'to form the side walls. The logs don't have to be all the same length, so long as they are long enough to wedge between these trees here.'

They all nodded in agreement and spread out into the forest. Finding the right logs and hauling them to the den-site was hard, tiring work. And it wasn't long before Drun and Trindle were spending more time with each other than looking for logs and hauling them back to the den! And there were murmurings among the rest of the group.

'Jimbo', called Conché, watching from her perch up in the trees, 'some of your friends are losing interest already.'

'I know,' replied Jimbo, 'but if we don't spend time building the walls strong enough, they won't be able to support the roof, so we've got to get it right even if it takes a lot of time.'

'You're right,' said Conché. 'I've watched plenty of others trying to build dens before – and what a mess they made!'

'Not this time!' cried Jimbo, as once more he wrapped his trunk around a log and began to haul.

Slowly – very slowly – the walls began to rise, and the shape of the den began to emerge – but at a price! They were all very, very tired and kept falling out with each other.

'I don't think it's fair!' said Jondle grumpily. 'Drun and Trindle have hardly done anything. I've had enough, I'm off home!'

Others murmured their agreement and followed Jondle.

'I know it's taken a long time,' pleaded Jimbo, to the few friends who had stayed behind, 'but unless we'd got the walls strong and firm, we'd have no chance of the roof staying up. So come on, let's get on with the roof.'

'"Let's get on with the roof,"' mimicked Conché, as she landed on Jimbo's head. 'And how exactly are we going to do that?'

'Oh, it's easy', said Jimbo, turning to his remaining friends to explain. 'We've enough room at this higher, open end to lift the logs up and allow them to roll down to the back. They'll be stopped at the far end by the uprights of the back wall.'

'But we're all so tired,' wailed his friends. 'We've not enough strength left to lift the logs so high.'

'We can do it', urged Jimbo, 'if we all lift together.'

So, one by one, with everyone giving their last ounce of energy, the logs were lifted into place and allowed to roll down to the back to make the roof. Unfortunately, not all the logs kept to course; some wobbled so far to one side that they even fell into the den, making it an even harder job to lift them out, reposition them, and lift them onto the roof once again.

After the third log had fallen into the den, Jimbo's few remaining friends had had enough. 'We can't do this anymore, Jimbo,' they said. 'Sorry, but we're going home.' And no amount of persuasion could stop them.

All alone now, Jimbo felt desolate. He sat down in the unfinished den and cried. And as if that weren't enough, the clouds darkened, and the rain poured down!

At first, Jimbo couldn't hear the little voice in his ear because of the sound of the rain. But slowly, the gentle voice crept into his heart, and he knew it was his beloved friend Conché. 'Come on, Jimbo', she whispered, 'you must finish the den.'

'Oh, no!' countered Jimbo. 'I'm too tired and wet, and the logs are too heavy. Besides, if I did finish it, it wouldn't only be me who would play in it. They would all want to as well.'

'But Jimbo', said Conché, 'can't you see – they can't finish the den, because –'

'Because nothing!' interrupted Jimbo. 'They're as capable as I am; they just can't be bothered!'

'No!' said Conché. 'It's true they're just as strong as you are, but they don't have your patience or persistence; that's what'll finish the den, Jimbo. They're great gifts, but not everybody is the same as you.'

Slowly, with a great deal of effort, Jimbo raised himself and started work once again. Before long the rain stopped and the sun broke through, and just before darkness fell, the den was finished.

It was true that in the years that followed, all the young elephants did play in the den, but it was never referred to as simply 'the den', but always as 'Jimbo's den'. As the story of its building was passed down the generations, it became a 'holy' place – a place which seemed to capture, in its very walls, the spirit of Jimbo: a spirit which inspired them all, in every generation, to keep going, no matter how hard the task, nor how long the road.

4
Jimbo's Whistling Nose

It was a terrifying sound! They had heard groaning and cracking all afternoon, but they reckoned they were the sounds to be expected from deep inside a mountain. But this latest sound was different, and they froze with fear. They held their breath as the rumble grew into an ear-splitting roar!

Clouds of dust billowed into their chamber, blocking out what little light they had. They trembled in concert with the shaking ground beneath their feet.

When the noise finally stopped and the dust began to settle, all they could hear in the eerie silence was the thumping of their own hearts. Suddenly the coughing began, a choking rasping cough, as they frantically tried to breathe again, and their minds and bodies struggled back to life.

Slowly the coughing ceased, and the silence returned. And into the silence, a whisper, an urgent pleading whisper out of the darkness: 'What was that? Are we trapped? Please don't tell me we're trapped?' And a second, reassuring whisper: 'It was a rockfall, I think. But don't worry; we'll soon find a way out.' But Jimbo knew that these were words of comfort rather than truth because he was pretty sure that the rocks which had tumbled down would almost certainly have blocked their passage and trapped them in the caves. Now he realized why, for generations, these caves had been called the 'forbidden' caves.

How the afternoon had changed! What was wrong with an exciting exploration of caves? Crawling through tunnels and squeezing between rocks? Daring one another to go into very dark places and feeling the surge of pride when the dare was won? But now look what had happened: they were trapped inside a mountain with little hope of getting out. 'And it's all my fault, I brought them here,' cried Jimbo quietly to himself.

Jimbo's friends were beginning to panic. Some were crying uncontrollably; some were stamping their feet in anger; and some were banging their heads on the roof as they reared up wildly on their hind legs. They were all very, very frightened.

Jimbo tried to calm them, but he couldn't make himself heard, so his friend Jondle shouted at the top of his voice: 'Quiet! Quiet! If we panic like this, it'll only make matters worse!' Gradually the elephants steadied themselves and, in the darkness, reached out to touch one another for comfort.

Jimbo quickly realized that they needed something to occupy their minds, otherwise, they would plunge headlong back into hysterical fear. He wracked his brains trying to think of something. Then he remembered that his father would always sing to him whenever he was afraid. So, he suggested they sing – to the weird sound of his whistling nose! Most of the time his friends made fun of his whistling nose. But his mother always said it was a 'special gift'. His nose didn't whistle all the time, only when he lifted it above his head. With a bit of practice, Jimbo had found that he could actually whistle a tune – of sorts!

So, Jimbo began to whistle, and Jondle urged them all to sing along as loudly as they could. The chamber reverberated to the sound, and Jimbo suddenly wondered if the singing wouldn't only soothe their nerves and calm their fears, but be heard outside the caves? With this in mind, Jimbo encouraged them all to sing at the top of their voices. Tune after tune he whistled, until they all began to lose their voices – and with them, in Jimbo's mind, their hope as well.

The darkness pressed down upon them like a smothering blanket, and their imaginations began to run wild. What was that brush on the shoulder? … Had a lion been sheltering in the darkness? … *Was it about to attack?* … What was that ear-splitting crack? … Was the whole mountain splitting apart? …

Jimbo had stopped whistling. He was barely holding himself together, but he knew he must; they were all relying on him to get them through. But how? Where was the rest of the herd? Would anyone even realise they were missing? And if they did, how would they know to come looking in the caves? Where was Conché? When had he last seen her? Was she with him when they had entered the caves? He couldn't remember. But what he was now sure of was that without some sign from them inside the caves, no one outside, including Conché, would know they were there. But what sign? It had to be a sound, a high-pitched sound, if it was to have any chance of penetrating the walls of their tomb. The only sound Jimbo could think of was, again, the whistle of his nose. He realized that he must whistle and whistle and whistle! And hope that its high pitch would penetrate the walls and be picked up by Conché, and she would bring help. So, he whistled and whistled …

The hours passed and Jimbo's whistle was almost gone – and he had long ago run out of fresh tunes, all he could do was whistle the same ones over and over again. His friends were now all huddled together with their heads hanging limply. Their fear and high tension had sapped all their energy. Their breathing had become laboured as the air began to run out. They were giving up hope. Jimbo was well aware that before long, the air would run out completely. So, with one huge last act of will, he began whistling again. Only one tune now – the tune Conché loved most.

Then they heard it! It was unlike any other sound they had heard in the caves before. It was the sound of rocks being flung aside! Before long they could hear elephants' voices calling to one another from the other side of the rockfall. His friends all pricked up their ears and began murmuring to one another. Was it too much to hope? Could it possibly be …?

Suddenly, a tiny speck of light appeared at the top of the fall, and a flicker of light as a little bird swept through! Then Jimbo heard that heavenly voice whispering in his ear, the voice of his beloved friend Conché: 'I'm so, so thankful for your whistling nose, Jimbo!'

5
Round Red Juicy Berries

'Conché', said Jimbo, as they strolled through the forest one sunny afternoon, 'have you ever heard of the "haunted glade"?'

'Oh, yes!' said Conché, 'I've heard of the haunted glade all right, and it's not somewhere you should ever go, Jimbo!'

'But from what I've heard,' said Jimbo, 'there's some beautiful round red juicy berries in the glade – in fact some of the biggest and juiciest in the whole of the forest!'

'Ah, yes, they're big and juicy all right,' said Conché 'but they won't do you any good.'

'Why not?' asked Jimbo. 'They're berries, aren't they? I'll bet they taste delicious, so why won't they do me any good?'

'Listen to me,' said Conché. 'They might taste delicious, but they're very, very bad for you. You should never ever touch them. In fact, you should never go into the glade in the first place. I've seen what can happen to elephants who go in there.'

'Hm, I'm afraid it's too late now', said Jimbo. 'I've arranged to go there this afternoon with some of my friends, to celebrate our New Year.'

'No, no, you mustn't go!' pleaded Conché, hopping agitatedly from one foot to the other on the top of Jimbo's head.

'But I've said I'd go', replied Jimbo defiantly, 'and I'm certainly not going to let my friends down, so don't try to stop me.' With that, Jimbo ran off very fast and headed straight for some thorn bushes, forcing Conché to jump for her life!

As Jimbo ran on, Conché kept her distance, circling above the trees with a growing dread filling her heart. Along the way Jimbo was joined by first one, then another of his friends, and by mid-afternoon they had reached their destination – the haunted glade.

They slowed as they approached the glade, and their mouths began to water as they caught their first glimpse of the huge round red juicy berries. What a feast they were in for! *But first things first*, thought Jimbo to himself. *It obviously isn't called the 'haunted glade' for nothing. So, what could it be?* he asked himself, while scanning the glade carefully. There were several elephants already in the glade – munching furiously, with juice streaming from the ends of their noses! Yum-yum! Apart from the elephants there was nothing else to be seen – except a few bones scattered here and there, picked clean by the usual scavengers. Well, perhaps there were more than a few – but that was to be expected with the number of animals such a glade would attract!

Suddenly the trees began to creak, and the leaves rustled in the swaying branches. Jimbo and his friends stopped dead in their tracks! Listening! Watching! They stood very still, daring hardly to breathe. Their eyes flicked furtively from tree to tree – but they could see nothing. Then, just as suddenly as it had arisen, the eddy wind died away. After a few minutes, they crept slowly forwards again towards the fruit-laden trees. With each step they were expecting 'unearthly' things to happen … but nothing did.

Soon, the big round red juicy berries were within their grasp – and they were even more delicious than they'd been told! They scooped up berry after berry, scrunching and swallowing, drenching their taste buds with the glorious nectar as it made its long journey into their stomachs!

From time-to-time Jimbo became aware that his stomach was swelling up. *Only to be expected*, he thought to himself, as he reached for the next berry … then the next … and the next. They were simply irresistible!

The sun was beginning to sink behind the trees when, deep within the recesses of his mind, Jimbo remembered the warning of Conché: 'I've seen what happens to the elephants that go into that glade ….' But, looking around, he couldn't see anything wrong. Everything seemed normal. Yet Conché was his very best friend, so what did she mean? He looked around again, but he couldn't see that any of them were coming to any harm – so he returned to the berries. But still the voice kept pushing into his mind: 'Please don't go in, Jimbo. I've seen what happens ….' Jimbo hesitated. The berries were so tempting, so delicious! *Oh, just one more,* he promised himself. Yet still the voice would not go away.

Eventually, putting all his faith in his friend, and with a huge act of will, Jimbo resisted his repetitive 'Just one more' response and backed slowly away from the berries to the edge of the glade.

The funny thing was that, when he turned to go back into the forest, the trees seemed to have moved! They seemed to have come closer together! How could that be? Trees couldn't just move like that! But they had, because when he tried to push through them to get out of the glade – he couldn't! Yet he knew that this was the way he had come in. And then the penny dropped – it wasn't the trees which had moved, it was him! His belly had swelled to an enormous size!!

Now he knew why it was called the 'haunted glade'. The temptation of the berries was like the work of an evil spirit: once you were addicted, there was no way out. Extremely bloated from eating so much fruit, you couldn't get through the trees to return to the forest!

Jimbo, with his heart beginning to pound and fear welling up inside him, cast madly around in his mind for a solution. Could it be that, just possibly, because of Conché's warning ringing persistently in his ears, he hadn't left it too late? Could it just be possible to squeeze his way out if he really pushed? He knew it would be very painful; he would probably lose a lot of skin! But what else could he do?

Over the years that followed, Jimbo could often be seen to stand very still, a shudder running through him, before a wistful smile played across his face. He was remembering his very sore and bruised body, as he had struggled with all his strength to force his way out of the glade. He was remembering, too, his friend Conché, dancing and singing deliriously in her favourite place on the top of his head and shouting her unforgettable words: 'Welcome home, Jimbo. I thought you were going to die, but you're alive!'

6
The Beautiful Dance

Jimbo was not feeling very well. He ached everywhere – his stomach ached, his lungs ached, his legs ached, even his nose ached! He didn't want to get up in the mornings, and when he did get up, he didn't want to do anything. Even his temper was short.

'Why don't you come with us to the river?' pleaded Trindle, timorously.

'Because I don't want to!' replied Jimbo sharply.

'Oh, come on,' chided Drun. 'You can't just mope about here all day: it's weeks since you last played with us.'

'Leave me alone,' said Jimbo grumpily. 'I don't want to play, and that's that!'

'But you don't want to do anything,' said Jondle.

'And what's that got to do with you?' Jimbo retorted. 'All of you just go away and leave me alone.'

Jimbo knew that something was wrong with him, but he'd no idea what it was. It wasn't something he had eaten – he didn't feel like eating much anyway. And it wasn't something that had anything to do with anyone else. *So, what's the matter with me?* he kept asking himself.

His mother thought it was because he had fallen out with Zuba, but he knew it wasn't that either. True, he hadn't seen her for quite some time: he could still remember the day she had walked away. For a long time, they had enjoyed each other's company: they had played together, walked together, and talked together. All his friends liked her – 'A match made in heaven,' they said.

But one day, during the third year of the last drought, a small group of elephants had decided to split from the herd and try their luck elsewhere. Zuba's family were part of that group. Jimbo's heart had broken as he watched her go. Tears were streaming from Zuba's eyes, and uncontrollable sobs were convulsing her chest. As she disappeared with her family into the blur of the rising heat, Jimbo had collapsed onto his knees and wept.

It was many months ago now, and though his heart still ached whenever he thought of her, he knew that life had to go on. No, his present malaise wasn't due to his parting with Zuba; rather, in a strange kind of way, he felt it was related to Conché, his best friend. He couldn't work out why, but he sensed that Conché wasn't *her* usual self, hadn't been for quite some time, and somehow it was affecting him.

Conché was aware of feeling down herself. She couldn't rid her mind of the memory of a wrong in her life which she should have put right. It went back a very long time – to the day she met Jimbo, in fact. All those years ago, and it had come back to trouble her now! She had tried and tried to put the matter behind her, but to no avail. Again and again, the memories kept flooding back, so she decided to leave Jimbo for a little while and fly away to her secret place.

Her secret place was in the High Trees, on the lower slopes of the Great Mountain. It was a place where she could be truly alone, with just the sound of her own heartbeat and the calls of the animals of the forest for company. But where, most of all, she would be calmed by the soothing voice of the wind in the high trees and the gentle rustle of the leaves … they never failed to speak to her deepest needs.

The first day came and went, but there was no wind. The second came, then the third … but still no wind. On the fourth day, she was feeling a little hungry – and very sad.

But then, into the depths of her sadness came the first stirring of the leaves, and slowly she became aware of the light, gentle breeze playing around her. Very slowly, oh so slowly, it soothed her aching heart. As the wind strengthened a little, she heard a whisper as if in the wind: 'You must learn to forgive, Conché; you may never forget, but you can learn to forgive … to forgive … to forgive.' The voice drifted away as the wind passed by. She was alone once more, on her perch in the top of the trees. But she felt different … she was aware of a peace deep within her … a peace which she had not felt for a very long time. And she knew that the rift with her friend, which had happened so long ago, was healed!

It had happened on the last day she had played with her friends on the Big Roller. She had been determined to win the game that day, but her overenthusiasm had led to recklessness. Diving too fast, she had been unable to swoop under the last branch. She had hit the branch at full speed, knocked herself unconscious, broken her wing, and ended up on the forest floor in extreme danger!

But worse, she learned later that her friends had left her … and her very best friend had told her parents that she was dead!

Luckily, she had been rescued from the forest floor by a wonderful elephant called Jimbo. He had lifted her onto the top of his head, where she stayed until her wing was healed and she was able to fly off and be united with her deliriously joyful parents.

But she found it hard to forgive her friends for leaving her, and particularly her best friend for causing her parents so much pain by telling them she was dead. Unfortunately, before she was able to make it up with her friend, her friend herself had died, after mistakenly eating a poison berry.

And it was this, the fact that she had not been able to forgive her friend before she died, which had weighed heavily on Conché's heart all these years and now, through the message of the wind of the high trees on the Great Mountain – 'You must learn to forgive … to forgive …' – the weight had finally lifted. Perhaps she could never forget, but she now knew how to forgive – and forgive not only her friend, but also herself.

She left her perch that evening and swept down to find her beloved Jimbo. As she perched on his head, between his huge flapping ears, and sang to him his favourite song, Jimbo's eyes sprang to life, and he began to dance a beautiful dance! He danced and danced until a new day dawned. Then he and Conché set off once more into the forest, with happiness in their hearts.

7
Don't Eat All the Worms

Recently, Jaco, the king of the elephants, had been noticing things that didn't seem quite right.

He had noticed, for instance, that in each of their feeding grounds, the trees weren't growing as well as they used to, and the fruit was not as big nor as juicy.

He had also noticed that their traditional waterholes were contracting in size, and some had even dried up altogether. And the rivers, they didn't seem to be running as fast as they used to.

So, what was wrong? What was happening? He knew he would have to do something pretty soon, or his herd would begin to starve, diseases would take hold, and the very young and the very old would die before their time.

He therefore called all his herd together and told them of his concerns. 'I am at a loss to know what is causing these changes, but I know we must do something and do it quickly, for the futures of our sons and daughters.'

There were murmurs among the herd as they reflected on what Jaco had said, and they all agreed that, come to think of it, they also had noticed things were not quite the same.

'You must all have noticed', Jaco continued, picking up on their conversations, 'how our elderly are growing frail much too early, and our young ones are not growing as fast and are not as robust as in earlier times. No doubt you've all got your own ideas about the causes of these things. But before we take any firm action, we must know for certain. So, I'm going to ask Jimbo to investigate, and let me have his findings and proposed solutions by the end of the second rains.'

Jimbo was taken by surprise and was very nervous about the task he had been given. 'Why me? Why me?' he pleaded to Conché, as the elephants dispersed, 'I've no idea where to start!'

'You underestimate yourself,' said Conché, feeling very proud of Jimbo having been chosen for such an important task. 'You've learnt a great many things in your life, Jimbo, and you have a very open and inquiring mind, so I'm sure you'll find the answers.'

'Oh, I don't know about that,' said Jimbo. 'Anyway, I've got no choice.'

Until the end of the second rains, Jimbo, assisted by his friends Trindle, Jondle, and Drun, together with the constant whispers of encouragement from Conché, worked hard to discover the real reasons for the serious changes taking place around them.

First, he looked at the trees, which parts of the trees were their main source of food? He discovered that with many of the trees, it was the bark – it was almost as tasty as the fruit and leaves, so they had taken to stripping the bark almost to the base of the trunk. This left the trees very exposed to all kinds of diseases, which in turn resulted in stunted growth and eventual death. In effect, they were killing off their own food supply.

The same applied to the young shoots and saplings. Tasty as they were, they provided little sustenance, and once they were eaten, all future supplies were lost.

Similarly, by eating fruit when it was too small, before it had ripened, however tempting this was when they were hungry, led to even greater hunger when their food source for the season was prematurely depleted.

There were other things too which Jimbo began to notice with his newly opened eyes. He was struck by the significant damage done by the simple passage of such a large number of elephants. So many shrubs, saplings, and young trees were inevitably trampled underfoot or broken to the ground. Once again, they were destroying their future resource.

He was also alarmed, when he took a much closer look, at the contraction of the waterholes and the reduced flow of the rivers. It had never struck Jimbo before just how little rain there had been in the last few years – or just how important water was to the elephants: without it they simply could not survive. Many of their 'wanderings', which he used to think they were when he was much younger, were actually planned migrations from one watering hole to another.

Jimbo and his friends discussed all these issues over and over again – but they found it very hard to come up with any solutions.

Then Conché remembered something her father used to say: 'Don't eat all the worms!' He used to apply it to just about everything – worms, bugs, berries – and now Conché was beginning to realize why.

She told Jimbo about it, explaining that, if they did not eat everything up, things could regenerate, ensuring something was always available, season after season. 'But how does that apply to us?' asked Jimbo. Trindle and Jondle looked at each other and shrugged – they had no idea! Drun simply pretended that he had not heard the question!

'We're going to have to put our thinking caps on, right now!' Jimbo declared. 'The king is expecting our report very soon.' So, they thought and thought – and thought. But nothing came!

'Okay,' said Jimbo growing exasperated, 'let's take each issue in turn and apply the words of Conché's father – "Don't eat all the worms" – as imaginatively as we can. It doesn't matter how ridiculous or far-fetched your ideas might be. You never know, one of them might just be our answer!'

Over the next few days, they hardly slept; each issue was exhaustively considered – until their brains hurt! But gradually the report came together.

At a great gathering, Jaco called upon Jimbo to present his report. As

Jimbo nervously stepped forward, Conché whispered in his ear: 'Just keep repeating my dad's words, Jimbo; they'll quickly understand.'

With renewed confidence Jimbo outlined the work they had done and the seriousness of the situation they were facing. Then he told them of the actions they were proposing:

'With regard to the diminishing food supply, we suggest three things:

- First – we should not eat bark from the trees below head height. This will allow the trees to recover for our next passage.
- Second – we should not eat the very early berries or the young saplings. This will allow them to grow to maturity, guaranteeing food for the future.
- Third – as we move through the land, we should split up into small groups, leaving twenty paces laterally between each group. This way, the untrodden areas will continue to flourish until our next passage.'

'Concerning our decreasing water supply, again we have three suggestions:

- First – we should deepen our existing waterholes. This will increase the volume of water they can hold, and so extend the supply of water for us through periods of drought.
- Second – we should construct a series of obstructions along the length of our rivers. This will reduce the rate of flow of water out of our territory.
- Third – during periods of drought, we must reduce all water activities. This will maximise the amount of water available for drinking.

'In conclusion, my king', continued Jimbo, 'may we suggest our herd adopt a slogan, practiced by many of our bird friends, which summarises the way we all should live our lives from now on, and something which we can pass on easily to future generations – "DON'T EAT ALL THE WORMS!"'

A murmur spread through the herd as they tried to understand the meaning of the slogan. Gradually the murmur grew into a thunderous roar which shook the forest: *'Don't eat all the worms!'* they bellowed to one another, over and over again.

Those standing close to the king, heard him shout into Jimbo's ear: 'That slogan you suggest, I like it, I like it very much indeed!'

8

'I Wish I Was Like You'

Jimbo was out walking alone in the forest. It was a beautiful day; the sun was shining, a gentle breeze was blowing, and the birds were singing their glorious tunes. Normally on such a day, Jimbo would have had a spring in his step; he would have been swinging his trunk and flapping his ears ... but not today. Today, his head was down, his trunk curled up, and he plodded very heavily.

Conché dropped down onto his head: 'What's the matter Jimbo?' she said. 'You look as if you've walked into a rock wall!'

'Leave me alone,' said Jimbo. 'I'm just fed up.'

'Ah,' said Conché, 'and you're fed up because –?'

'Just because!' growled Jimbo.

'Not good enough!' said Conché.

'All right then,' said Jimbo. 'I'm fed up because I'm not as big and strong as the other elephants: I can't bend the bigger trees to get to the best leaves and fruit; I can't pull the heavier logs to build dams and shelters; I can't run as fast as they can – and ... and I'm the one who's always left to clear up the mess ... and –'

'Oh dear,' said Conché, 'we are feeling sorry for ourselves today, aren't we! Have you listened to yourself, Jimbo? Have you ever stopped to think about things for a moment? Don't you think you're exaggerating a little? What about me for instance?'

'What about you?' replied Jimbo grumpily.

'Well, compared to you I'm nothing!' continued Conché. 'What can I lift? – the most I can build is a very fragile nest! What trees can I bend? For me, trees are dangerous obstacles at every turn – but you know that, don't you, because that's how we met, when I broke my wing on Big Roller! And what about when it rains, and the wind blows hard? They don't trouble you, but for me, the wind blows me all over the place, and the rain feels like boulders pummelling my body!'

'Maybe,' reflected Jimbo, 'but I can't sing like you. When you sing, everything in the forest stops to listen, it's so beautiful. But when I sing, they all clamp their ears and squeeze their eyes shut! And look at this stupid nose! Have you ever seen anything like it!? And it whistles, you know; you know that don't you?'

'Of course, I know that Jimbo, but you seem to have forgotten how it saved your life once – and those of your friends – when the roof collapsed in the caves!'

Suddenly, a swooshing noise filled the air, and a huge net sprang up all around them! Everything was a blur; they were ensnared in the net! Jimbo didn't even know which way was up! His first reaction was to panic!

He thrashed about wildly – but it only made matters worse. Whichever way he turned, the net tightened.

Now he was angry. He started thumping the net with his great big feet and shaking his head vigorously to free his tusks, which were entangled in the net. No one was going to capture *him* like this! But the net grew evermore tangled around him.

Conché, meanwhile, had overcome her immediate confusion and was able to fly raggedly through the net and escape to the safety of the trees. From there, as she anxiously watched Jimbo's frantic efforts to escape, she tried to work out what had happened.

Her swift-moving eyes saw that the net was being held together at the top, just above Jimbo's head. From this Conché reasoned that each of the corners must have been tied to the top of their own slender tree, and each of these trees must have been bent backwards and held down with a central trip mechanism. The mechanism would be activated whenever an animal stepped into the concealed net.

After careful thought, Conché worked out that if Jimbo was to push his very long nose through the net, he could grab one of the holding trees, and, using all his weight and strength, break the tree, making a gap big enough for him to scramble to safety.

Quickly, Conché flew down into the net and onto the top of Jimbo's head. There she shouted her idea into Jimbo's ear.

Within minutes, Jimbo had brought one of the trees holding the net crashing down and was able to escape long before the poachers arrived.

'You see,' said Conché to Jimbo, as they broke out of the forest and felt the welcoming greeting of a warm breeze, 'by pooling our abilities, however small and different, we can conquer the world! If I hadn't been *me* and you hadn't been *you*, we would still have been caught in that net!'

APPENDIX 1

List of Stories, Poems, Songs, Proverbs, and
Giants of the Twentieth Century

1. Love Changes Everything
2. Tribute to John Keats
3. First Letter of St John
4. Selection of Biblical Passages concerning Living in the Light
5. The Small Beatitudes
6. The Shepherd-Boy and the Giant Warrior
7. The Shepherd-Boy and the Prince
8. The Mother and Daughter-in-Law
9. Romeo and Juliet
10. The Merchant of Venice
11. Mahatma Gandhi
12. Martin Luther King
13. Nelson Mandela
14. The Good Samaritan
15. The Lost Son
16. The High-Ranking Religious Professional and the Looked-Down-Upon Tax Collector, at Prayer
17. The Unforgiving Servant
18. The Resourceful Servant
19. The Farmer Sowing His Seed
20. Pithy Sayings of Jesus
21. 'Do It Anyway'
22. Well-Known Proverbs
23. The Story of Abraham

APPENDIX 2

Stories for Children … The Jimbo and Conché Stories

1. A Nose for Someone in Trouble
2. Stuck in a Hole
3. The Log Den
4. Jimbo's Whistling Nose
5. Round Red Juicy Berries
6. The Beautiful Dance
7. Don't Eat All the Worms
8. I Wish I Was Like You

REFERENCED BIBLIOGRAPHY

Armstrong, Karen. *A History of God*. Vintage 1999.

British Social Attitudes Survey 2018.

Brown, D.E. *Human Universals*. McGraw-Hill 1991.

Curry, Oliver Scott. *Darwin Day Lecture 2021*. Humanist UK.

Dalai Lama and Howard C. Cutler. *The Art of Happiness*. Hodder and Stoughton 1999.

Freire, Paulo. *Cultural Action for Freedom*. Penguin Books 1975.

Gandhi, M.K. *Collected Works* 1958 onwards.

Geldof, Sir Bob. *A Tribute to the Poet John Keats*. The *I* February 2018.

Goleman, Daniel. *Emotional Intelligence*. Bloomsbury 1996.

Harris, Sam. *The Moral Landscape*. Black Swan 2010.

—— *Waking Up*. Black Swan 2014.

King, Martin Luther. *Strength to Love*. Fontana Books 1987.

Mandela, Nelson. *Long Walk to Freedom*. Abacus 1995.

Maslow, Abraham. *Motivation and Personality*. Harper 1954.

Peterson, Eugene H. *The Message* (the Bible in Contemporary Language). Nav Press 2003

Pietsching, J., and M. Voracek. "One Century of Global IQ Gains: A Formal Meta-analysis of the Flynn Effect (1909-2013)." *Perspectives on Psychological Science,10.* 2015.

Pinker, Steven. *Enlightenment Now.* Penguin Books 2018.

—— *The Blank Slate.* Penguin Books 2019.

Scott, R. A. *Miracle Cures: Saints, Pilgrimage, and the Healing Powers of Belief.* Berkeley: University of California Press 2010.

Wilson, Bill. *Climbing the Stairway to Heaven.* Farthings Publishing 2016.

—— *Faith Refractioned.* Farthings Publishing 2020.

Wootton, D. *The Invention of Science: A New History of the Scientific Revolution.* Harper Collins 2010.

Zahn-Wexler, et al. "Development of Concern for Others." *Developmental Psychology*, 28. 1992.

ABOUT THE AUTHOR

Bill Wilson, a native of County Durham, now lives in Lincolnshire. He has studied civil engineering, theology, philosophy, organization theory, behavioral science, and education.

He has practiced professionally as a civil engineer, specializing in bridge design and construction; as a minister of religion; as a lecturer, specializing in industrial relations; and as principal of a further and higher education college. He is also the author of *Faith Refractioned*.